Problem Books in Mathematics

---

Edited by K. Bencsáth
P.R. Halmos

Springer
*New York*
*Berlin*
*Heidelberg*
*Barcelona*
*Hong Kong*
*London*
*Milan*
*Paris*
*Singapore*
*Tokyo*

## Problem Books in Mathematics

Series Editors: K. Bencsáth and P.R. Halmos

**Polynomials**
by *Edward J. Barbeau*

**Problems in Geometry**
by *Marcel Berger, Pierre Pansu, Jean-Pic Berry, and Xavier Saint-Raymond*

**Problem Book for First Year Calculus**
by *George W. Bluman*

**Exercises in Probability**
by *T. Cacoullos*

**Probability Through Problems**
by *Marek Capiński and Tomasz Zastawniak*

**An Introduction to Hilbert Space and Quantum Logic**
by *David W. Cohen*

**Unsolved Problems in Geometry**
by *Hallard T. Croft, Kenneth J. Falconer, and Richard K. Guy*

**Berkeley Problems in Mathematics**
by *Paulo Ney de Souza and Jorge-Nuno Silva*

**Problem-Solving Strategies**
by *Arthur Engel*

**Problems in Analysis**
by *Bernard R. Gelbaum*

**Problems in Real and Complex Analysis**
by *Bernard R. Gelbaum*

**Theorems and Counterexamples in Mathematics**
by *Bernard R. Gelbaum and John M.H. Olmsted*

**Exercises in Integration**
by *Claude George*

**Algebraic Logic**
by *S.G. Gindikin*

**Unsolved Problems in Number Theory** (2nd ed.)
by *Richard K. Guy*

*(continued after index)*

Marek Capiński   Tomasz Zastawniak

# Probability Through Problems

With 27 Figures

Marek Capiński
Department of Finance
Nowy Sacz Graduate School of
Business—NLU
Zielona 27
33-300 Nowy Sącz
Poland
capinski@sigma-wsb-nlu.edu.pl

Tomasz Zastawniak
Department of Mathematics
University of Hull
Kingston upon Hull HU6 7RX
England
T.J.Zastawniak@maths.hull.ac.uk



Mathematics Subject Classification (2000): 60-01, 00A07

Library of Congress Cataloging-in-Publication Data
Capiński, Marek, 1951–
Probability through problems / Marek Capiński, Tomasz Zastawniak.
p. cm. — (Problem books in mathematics)
Includes bibliographical references and index.
ISBN 0-387-95063-X (alk. paper)
1. Probability—Problems, exercises, etc. I. Zastawniak, Tomasz, 1959– II. Title.
III. Series.
QA273.25 .C36 2000
519.2′076—dc21 00-056272

Printed on acid-free paper.



Production managed by MaryAnn Brickner; manufacturing supervised by Jerome Basma.
Photocomposed copy prepared using the authors' LaTeX2e files.
Printed and bound by Maple-Vail Book Manufacturing Group, York, PA.
Printed in the United States of America.

9 8 7 6 5 4 3 2 1

ISBN 0-387-95063-X SPIN 10770576

Springer-Verlag New York Berlin Heidelberg
*A member of BertelsmannSpringer Science+Business Media GmbH*

# Preface

This book of problems has been designed to accompany an undergraduate course in probability. It will also be useful for students with interest in probability who wish to study on their own.

The only prerequisite is basic algebra and calculus. This includes some elementary experience in set theory, sequences and series, functions of one variable, and their derivatives. Familiarity with integrals would be a bonus. A brief survey of terminology and notation in set theory and calculus is provided.

Each chapter is divided into three parts: Problems, Hints, and Solutions. To make the book reasonably self-contained, all problem sections include expository material. Definitions and statements of important results are interlaced with relevant problems. The latter have been selected to motivate abstract definitions by concrete examples and to lead in manageable steps toward general results, as well as to provide exercises based on the issues and techniques introduced in each chapter.

The hint sections are an important part of the book, designed to guide the reader in an informal manner. This makes *Probability Through Problems* particularly useful for self-study and can also be of help in tutorials. Those who seek mathematical precision will find it in the worked solutions provided. However, students are strongly advised to consult the hints prior to looking at the solutions, and, first of all, to try to solve each problem on their own.

Hints are given for all problems, the majority of which also have fully worked solutions. To avoid repetition, we have left out a few solutions that are very similar to the preceding ones. Important items such as definitions, theorems, and some problems of theoretical nature are marked with ▶. Hard problems, which can be omitted on first reading, are designated by $*$, and those without solution by $\circ$.

The book begins with a motivating chapter on modeling random experiments. This is followed by Chapters 2 through 6 devoted to the mathematical structures underpinning probability theory: classical probability spaces and related combinatorial problems, fields of sets (also known as algebras), finitely additive probability, sigma fields (sigma algebras), and countably additive probability. Chapter 7 is concerned with the crucial notions of conditional probability and independence. Random variables and their probability distributions are discussed in Chapter 8. The notions of mathematical expectation and variance are studied in Chapter 9, followed by a careful step-by-step exposition of conditional expectation in Chapter 10. Characteristic functions are the subject of Chapter 11. Chapter 12 offers problems designed to illustrate the law of large numbers and the central limit theorem, the emphasis being on understanding the consequences and applications. The Bibliography contains a list of books recommended either as background reading or to consolidate and supplement this course of study.

Marek Capiński
Kraków, Poland

Tomasz Zastawniak
Kingston upon Hull, England

August 20, 2000

# Contents

# Terminology and Notation

This section introduces some standard symbols and terminology, mainly in set theory, to be used throughout this book. Notation concerned with numbers, functions, and basic notions in calculus is also collected here.

- $x \in A$ means that an element $x$ *belongs* to a set $A$;

  $x \notin A$ means that $x$ *does not belong* to $A$.

- $\emptyset$ denotes the *empty set*; it has no elements; that is, $x \notin \emptyset$ for any $x$.

- $\{x_1, x_2, \ldots, x_n\}$ or $\{x_i\}_{i=1}^{n}$ denotes a *finite set* consisting of elements $x_1, x_2, \ldots, x_n$; the empty set $\emptyset$ will also be regarded as a finite set.

  $\#A$ is the *number of elements* of a finite set $A$; for example, $\#\emptyset = 0$ and $\#\{0, 1, 2\} = 3$.

  We say that a set is *infinite* whenever it is not finite;

  $\{x_1, x_2, \ldots\}$ or $\{x_i\}_{i=1}^{\infty}$ denotes a *countable set*, consisting of a sequence of elements $x_1, x_2, \ldots$; any finite set, including the empty set $\emptyset$, will also be regarded as a countable set.

  $\{x \in A : P(x)\}$ stands for the set of elements $x$ belonging to $A$ that satisfy a certain property $P$; for example, $\{n \in \mathbb{N} : 2 < n < 6\} = \{3, 4, 5\}$.

- $A \subset B$ denotes the *inclusion of sets*; it means that $A$ is a *subset* of $B$; that is to say, $x \in A$ implies $x \in B$.

  $2^A$ denotes the set consisting of all subsets of $A$; it is called the *power set* of $A$.

- $A \cup B$ is the *union* of sets $A$ and $B$; it consists of all elements $x$ such that $x \in A$ or $x \in B$.

  For a sequence of sets $A_1, A_2, \ldots$, the *union*

  $$\bigcup_{n=1}^{\infty} A_n = A_1 \cup A_2 \cup \cdots$$

  consists of all elements $x$ such that $x \in A_n$ for some $n = 1, 2, \ldots$ .

  More generally, for an arbitrary family of sets $\mathcal{A}$, the *union* $\bigcup \mathcal{A}$ consists of all elements $x$ such that $x \in A$ for some set $A$ in $\mathcal{A}$.

- $A \cap B$ is the *intersection* of sets $A$ and $B$; it consists of all elements $x$ such that $x \in A$ and $x \in B$.

  The *intersection*

  $$\bigcap_{n=1}^{\infty} A_n = A_1 \cap A_2 \cap \cdots$$

  of a sequence of sets $A_1, A_2, \ldots$ consists of all elements $x$ such that $x \in A_n$ for all $n = 1, 2, \ldots$ .

  More generally, the *intersection* $\bigcap \mathcal{A}$ of a family of sets $\mathcal{A}$ consists of all elements $x$ such that $x \in A$ for all sets $A$ in $\mathcal{A}$.

- $A \setminus B$ is the *difference* of sets $A$ and $B$ or, in other words, the *complement* of $B$ in $A$; it consists of all elements $x$ such that $x \in A$ and $x \notin B$.

  $A \triangle B = (A \setminus B) \cup (B \setminus A)$ is the *symmetric difference* of sets $A$ and $B$; an element $x$ belongs to $A \triangle B$ whenever $x \in A \cup B$ but $x \notin A \cap B$.

- $A \times B$, the *Cartesian product* (or simply *product*) of sets $A$ and $B$, consists of all ordered pairs $(x, y)$ such that $x \in A$ and $y \in B$.

  The product $A_1 \times \cdots \times A_n$ of $n$ sets $A_1, \ldots, A_n$ consists of ordered $n$-tuples $(x_1, \ldots, x_n)$ such that $x_i \in A_i$ for each $i = 1, \ldots, n$.

  In particular, $A^n = A \times \cdots \times A$ is the product of $n$ copies of a set $A$.

- Sets $A$ and $B$ are said to be *disjoint* if $A \cap B = \emptyset$.

  The sets in a family $\mathcal{A}$ are said to be *pairwise disjoint* if $A \cap B = \emptyset$ for any $A, B$ in $\mathcal{A}$ such that $A \neq B$.

- The following sets of numbers will be used:

  $\mathbb{N} = \{1, 2, 3, \ldots\}$, the set of *natural numbers*;

  $\mathbb{Z} = \{\ldots, -2, -1, 0, 1, 2, \ldots\}$, the set of *whole numbers* or *integers*;

$\mathbb{Q} = \{\frac{a}{b} : a, b \in \mathbb{Z}$ and $b \neq 0\}$, the set of *rational numbers*;

$\mathbb{R}$, the set or *real numbers*;

$\mathbb{C} = \{x + iy : x, y \in \mathbb{R}\}$, the set of *complex numbers*, where $i = \sqrt{-1}$ is the *imaginary unit*.

*Intervals* $(a, b)$, $[a, b)$, $(a, b]$, $[a, b]$ in $\mathbb{R}$ with endpoints $a, b \in \mathbb{R}$; these intervals consist of numbers $x \in \mathbb{R}$ such that, respectively, $a < x < b$, $a \leqslant x < b$, $a < x \leqslant b$, $a \leqslant x \leqslant b$; for intervals of the form $(a, b)$ or $(a, b]$, we also allow $a$ to be $-\infty$; for intervals of the form $(a, b)$ or $[a, b)$, we allow $b$ to be $+\infty$.

$\sup A$ is the *supremum* or the *least upper bound* of a set $A \subset \mathbb{R}$.

$\inf A$ is the *infimum* or the *greatest lower bound* of a set $A \subset \mathbb{R}$.

- $f : A \to B$ denotes a *function* from a set $A$ into a set $B$.

  If $b \in B$ is the value of $f$ at $a \in A$, we write $f : A \ni a \mapsto b \in B$ or $f : a \mapsto b$ or simply $f(a) = b$.

  If $f$ is a one-to-one function, its *inverse* is written as $f^{-1}$.

  The *image* of a set $C \subset A$ under $f$ is denoted by $f(C)$; it consists of all elements of the form $f(x)$, where $x \in C$.

  The *inverse image* of a set $D \subset B$ under $f$ is denoted by $f^{-1}(D)$; it consists of all elements $x \in A$ such that $f(x) \in D$.

  In probability theory, the inverse image is often denoted by $\{f \in D\}$ instead of $f^{-1}(D)$.

  $\mathbb{I}_A$ is the *indicator function* of a set $A$ defined as follows: $\mathbb{I}_A(x) = 1$ if $x \in A$, and $\mathbb{I}_A(x) = 0$ if $x \notin A$.

- If a sequence of numbers $a_1, a_2, \ldots$ has a *limit* $a$ as $n$ tends to $\infty$, this will be denoted by $\lim_{n\to\infty} a_n = a$ or simply by $\lim a_n = a$ or by writing $a_n \to a$ as $n \to \infty$; if, in addition, $a_n$ is a monotone sequence, that is, either a non-decreasing or a non-increasing sequence, then we shall write $a_n \nearrow a$ or $a_n \searrow a$, respectively.

  The sum of a convergent series with terms $a_1, a_2, \ldots$ will be written as $\sum_{k=1}^{\infty} a_k$ or $a_1 + a_2 + \cdots$; it is defined to be the limit of the sequence with terms $\sum_{k=1}^{n} a_k = a_1 + a_2 + \cdots + a_n$ as $n \to \infty$.

  For a function $f$ from $\mathbb{R}$ into $\mathbb{R}$, if $f(x)$ has a *limit* $a$ as $x$ tends to $b$, it will be denoted by $\lim_{x\to b} f(x) = a$; alternatively, we shall write $f(x) \to a$ as $x \to b$; if the limit as $x$ approaches $b$ either from the left or from the right is considered, we shall write $x \to b^-$ or, respectively, $x \to b^+$ in these formulas; if the convergence is monotone, that is, $f$ is either non-decreasing or non-increasing, then we shall write $f(x) \nearrow a$ or $f(x) \searrow a$.

- The *derivative* of a function $f$ at a point $a$ is written as $f'(a)$ or $\frac{df}{dx}(a)$ or $\left.\frac{df(x)}{dx}\right|_{x=a}$; the *derivative function* is denoted by $f'$ or $\frac{df}{dx}$.

  The *second derivative* of $f$ is denoted by $f''$ or $\frac{d^2 f}{dx^2}$; the $n$th *derivative* is denoted by $f^{(n)}$ or $\frac{d^n f}{dx^n}$.

- The *antiderivative* or the *indefinite integral* of a function $f$ is denoted by $\int f(x)\,dx$; it is a function $F$ whose derivative is equal to $f$, that is, $F' = f$.

  For a function $f$ defined on a closed interval $[a, b]$, the *definite integral* of $f$ between $a$ and $b$ is denoted by $\int_a^b f(x)\,dx$.

  The same symbol $\int_a^b f(x)\,dx$ is used to designate the *improper integral* of $f$ defined on an open-ended interval $(a, b]$ or $[a, b)$ or $(a, b)$, the open end on the left being allowed to be $-\infty$ and that on the right $+\infty$.

# 1
# Modeling Random Experiments

## 1.1 Theory and Problems

Suppose we have an experiment (phenomenon, trial) with random outcomes such as, for example, the game of roulette (outcome: a number between 0 and 33) or a journey by train (outcome: the delay of arrival at the destination measured in minutes). We may be vitally interested in answers to questions such as:

How likely is it that the number in roulette is even?
How likely is it that the delay will not exceed 20 minutes?

The mathematical investigation of such questions begins after we find a suitable mathematical model of the random situation. Only then can the questions be addressed rigorously within the model.

We first choose a set $\Omega$ consisting of all possible outcomes of the trial. This set will be called a *probability space*. It is natural to take

$$\Omega_{\text{roulette}} = \{0, 1, \dots, 33\} \quad \text{in the case of roulette,}$$
$$\Omega_{\text{delay}} = [0, \infty) \quad \text{in the case of train journey.}$$

(We prefer the unbounded interval to avoid answering the question: What is the maximum possible train delay? In addition, we have implicitly assumed that the train is certain to arrive eventually at its destination by not allowing the delay to be $\infty$.)

Each statement of the form

The number in roulette is even.
The train is no more than 20 minutes late.

corresponds in a natural way to a subset of the set of all outcomes:

$$\{2, 4, \ldots, 32\} \subset \Omega_{\text{roulette}}$$

(in roulette, 0 is considered neither even nor odd),

$$[0, 20] \subset \Omega_{\text{delay}}.$$

Such sets are called *events.* To make the whole theory work, later we shall impose some conditions on families of events. First, we want the complement of any event to be an event (taking the complement of a set corresponds to negating the corresponding statement). Second, larger events can be produced from smaller ones by taking their union (taking the union of sets corresponds to joining the statements with the word "or"). For example, the same event

$$\Omega_{\text{roulette}} \setminus \{2, 4, \ldots, 32\} = \{0\} \cup \{1, 3, \ldots, 33\}$$

is defined by the statements

The number in roulette is not even.
The number in roulette is zero or odd.

We say that an event $A$ *takes place* if, once the experiment is executed, the actual outcome belongs to $A$.

It is customary to assume that the answer to the question

How likely is it that the event $A$ takes place?

is a number between 0 and 1. This number is called the *probability* of the event $A$ and is denoted by $P(A)$, so $P$ is a function assigning numbers to events:

$$P : A \mapsto P(A) \in [0, 1].$$

It is called a *probability measure.* We always assume that $P(\Omega) = 1$. This means that the probability of an event that is certain to happen is 1.

A fundamental difficulty in dealing with a function of this kind is that it is defined on sets rather than on numbers. However, we are familiar with some set functions such as length, area, or volume. The crucial feature common to all of these is this: To measure a set (its length, area, or volume), we can decompose it into smaller parts, measure each of them, and add the results. We shall assume that probability measures enjoy the same property, which is called *additivity.*

Now we are going to discuss the following basic question: How do we choose the set of outcomes $\Omega$ and how do we define a suitable probability measure $P$ for a given random situation? Any rigorous mathematical activity may begin only after a mathematical model consisting of $\Omega$ and $P$ has been chosen.

It is important to point out that the choice of $\Omega$ and $P$ is not unique and may be a matter of some dispute. Here, we shall try to make a choice based on our intuition, which may, for example, rely on some physical symmetries of the objects involved, our experience, or common sense.

For example, most people agree that if we toss a coin, there are two possible outcomes: heads or tails with probability $\frac{1}{2}$ each. Most people also believe that in a long series of tosses, the results will be divided approximately fifty-fifty. In fact few people have ever thrown a coin more than a dozen times. Our belief takes into account the regularity of the coin's shape (which is often expressed by saying that the coin is "fair"). However, if you examine a coin closely, you will see that it is never regular. If the heads side were bulging out a bit more, the above choice might not be correct. If the coin were thick enough, we would have to admit another possible outcome: landing on the edge.

Sometimes, the choice may result from a very pragmatic approach: If we want to be able to compute anything at all, we have to choose a simple model. Usually, this is the case if $\Omega$ has infinitely many elements. For example, if $\Omega = [0, +\infty)$, we may tend to assume that $P$ should be given by an explicit and simple formula.

Models based on a long series of experiments or on samples taken from a large population belong to a branch of mathematics called *statistics.*

**Example 1.1** Three coins are tossed: two nickels and a dime. We pay 10¢ to enter the game and receive those coins which fall heads up. What is the probability of winning some money?

*Solution* We first list some possible choices of the set of outcomes:

1. $\{0, 5, 10, 15, 20\}$, the amounts shown;
2. $\{-10, -5, 0, 5, 10\}$, our loss or gain, taking into account the fee paid;
3. $\{L, D, W\}$, game lost, drawn, or won, respectively.

The first two choices are, in fact, equivalent. The third neglects the actual amount won or lost.

Now we face the problem of selecting the measure. All we need is to prescribe the values of $P$ on one-element sets for each choice of $\Omega$. The probabilities of events can be computed using the additivity principle. The simplest choice of $\frac{1}{5}$ in the first two cases and $\frac{1}{3}$ in the third leads to a contradiction: The probability of winning the game would be $\frac{2}{5}$ and $\frac{1}{3}$, respectively. At this stage, it is difficult to say what the appropriate numbers should be. Therefore, we have to continue the quest for an $\Omega$ that would allow us to agree on the probability measure. We mark the coins so that we can distinguish between the two nickels. The 5¢ coins are given numbers 1 and 2 and the 10¢ coin gets number 3. All possible outcomes can be written as follows:

$$\Omega = \{(1H, 2H, 3H), (1H, 2H, 3T), (1H, 2T, 3H), (1H, 2T, 3T), \\ (1T, 2H, 3H), (1T, 2H, 3T), (1T, 2T, 3H), (1T, 2T, 3T)\}.$$

It seems reasonable to put $P(\{\omega\}) = \frac{1}{8}$, where $\omega$ is any element of $\Omega$, since each of the outcomes is equally likely if the coins are fair. Now we can perform computations assigning probabilities to the subsets of $\Omega$. We employ the basic property that the measure of the union of disjoint pieces is the sum of their measures. So the probability of winning is $\frac{3}{8}$, of drawing is $\frac{2}{8}$, and of losing is $\frac{3}{8}$.

At this stage, we can see that the three sets of outcomes proposed earlier can be equipped with appropriate probability measures in the following way:

1. $\Omega_1 = \{0, 5, 10, 15, 20\}$,
   $P(\{0\}) = P(\{20\}) = \frac{1}{8}$, $P(\{5\}) = P(\{10\}) = P(\{15\}) = \frac{2}{8}$;
2. $\Omega_2 = \{-10, -5, 0, 5, 10\}$,
   $P(\{-10\}) = P(\{10\}) = \frac{1}{8}$, $P(\{-5\}) = P(\{0\}) = P(\{5\}) = \frac{2}{8}$;
3. $\Omega_3 = \{L, D, W\}$,
   $P(\{L\}) = P(\{W\}) = \frac{3}{8}$, $P(\{D\}) = \frac{2}{8}$.

**Example 1.2** Two students have an appointment with their tutor at 12 o'clock. The first student makes an effort to be on time, but he lives far from the university and a recent 1-inch snowfall in Oxfordshire has paralyzed the county. He may be late, but will certainly make it before 1 p.m. The time of his arrival depends on so many unpredictable factors that we regard it as random. The other student lives next to the university, but he forgot about the appointment and may come at any time after having remembered it.

*Solution* We take $\Omega = [0, 1]$, representing the delay measured in hours. By events, we mean various intervals. For example, $A = [0, \frac{1}{2}]$ corresponds to the fact that a student arrived no later than 12:30. To measure intervals, we introduce an auxiliary function $f$ (called a *density*) and then $P(A)$ is the area under the graph of $f$ above the interval $A$. For the first student, we take $f$ to be $f(x) = -2x + 2$ for $x \in [0, 1]$, see Figure 1.1.

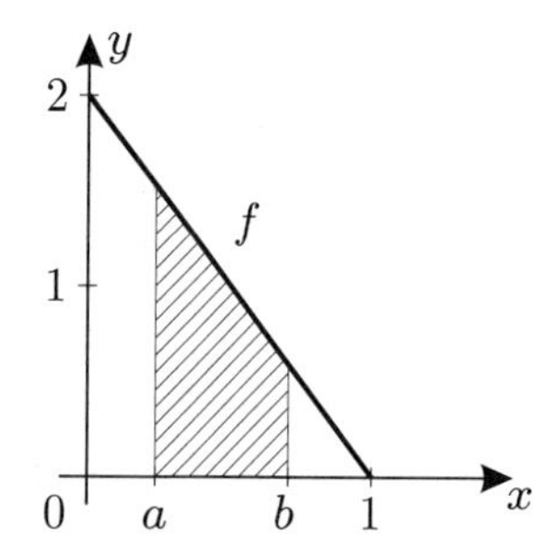


FIGURE 1.1. Density $f$ for the first student

Having chosen this measure, it is easy to see that, for example, $P([0, \frac{1}{2}]) = \frac{3}{4}$, while $P([\frac{1}{2}, 1]) = \frac{1}{4}$. This choice reflects our belief that it is more likely that the student will arrive earlier rather than later. Of course, it is a very rough approximation of complex reality. For instance, it may be more realistic to take a function with peaks at 0, $\frac{1}{3}$, and $\frac{2}{3}$, which may correspond to the bus timetable.

The absent-mindedness of the second student can be described by taking $f(x) = 1$ for all $x \in [0, 1]$, see Figure 1.2. This is called the *uniform density.*

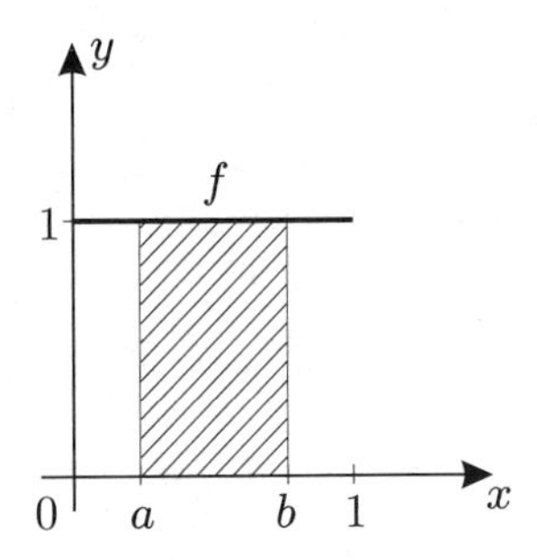


FIGURE 1.2. Density $f$ for the second student

The probability related to this function is given by $P([a, b]) = b - a$ and we call it the *uniform probability.*

Now try to propose the sets of outcomes and the measures in the following situations. If your answers are similar to the solutions below, then this only means that your way of thinking is close to ours, since no such thing as one correct solution exists here.

**1.1** A drawing pin is tossed.

**1.2** A coin is tossed twice.

**1.3** A die is rolled. What about two dice?

**1.4** A die and a coin are thrown simultaneously.

**1.5** A mug falls down from the table (luckily, an empty one).

**1.6** From a pack of 52 cards we draw 2.

**1.7** A pack of six numbered cards is shuffled and the numbers are revealed one by one.

**1.8** We play a series of chess games. The winner is the one who first scores three points, where one point is obtained for a single win and draws do not count.

**1.9** We keep throwing a coin until it lands heads up.

**1.10** The temperature outdoors is measured.

**1.11** How long shall we wait for a bus at a stop?

**1.12** How many days does a letter posted in Oxford take to reach London?

**1.13** What is the number of misprints on this page?

**1.14** How many road accidents happened today in your city?

**1.15** What time does your watch show now?

## 1.2 Hints

**1.1** The pin may land point up or down. Do you think these cases are equally likely?

**1.2** This is the same as if we tossed two coins once. Then, the outcome is a pair $(a, b)$, where $a$ and $b$ may be tails or heads.

**1.3** Assuming that the die is regular (fair), we have six equally likely outcomes. For two dice, an outcome is a pair $(a, b)$, where $a$ is the number shown on the first die and $b$ on the second.

**1.4** Again, we have pairs $(a, b)$ here, but we have to be careful, since the meaning and the range of $a$ and $b$ are different.

**1.5** In this example, there are many possibilities: We may want to know whether the mug breaks or not, into how many pieces, which side up it lands (if in one piece).

**1.6** The description of the experiment is not precise: We can draw two cards at once or draw one, return it to the pack, and draw again.

**1.7** The result can be written as a sequence of six numbers.

**1.8** What is the maximum length of the match? How many games may be played?

**1.9** Is there a number $N$ such that in $N$ throws we shall certainly have heads up?

**1.10** We need to choose the scale and then decide what accuracy is possible.

**1.11** Is this a bus stop in the first place (in some cities they have trams)? Does the bus have a timetable?

**1.12** Do we include the possibility that it will never arrive?

**1.13** In the extreme case, each letter on page 10 may be a misprint. (We certainly hope it is not so.) Thus, there are at most finitely many misprints.

**1.14** The number of accidents is not necessarily limited by the number of vehicles in the city. A vehicle can be involved in more than one accident on the same day.

**1.15** Depends on the type of display and the individual working habits.

## 1.3 Solutions

**1.1** Since the pin may land point up or down, we take $\Omega = \{U, D\}$. One may feel that the outcome $D$ is more likely and put, for example, $P(\{D\}) = 0.75$ and, consequently, $P(\{U\}) = 0.25$. However, a more elaborate analysis of the shape of the pin is required, or we have to perform a long series of experiments and take the average frequency as the probability. The geometric argument is that if you imagine that the pin is circumscribed by a sphere, the majority of the points on the sphere correspond to $D$.

**1.2** $\Omega = \{(H,H), (H,T), (T,H), (T,T)\}$, $T$ stands for tails, $H$ for heads, $P(\{\omega\}) = \frac{1}{4}$ for each $\omega \in \Omega$.

**1.3** For one die: $\Omega = \{i : i = 1, \dots, 6\}$ with $P(\{\omega\}) = \frac{1}{6}$ for each $\omega \in \Omega$. For two dice: $\Omega = \{(i,j) : i,j = 1, \dots, 6\}$ with $P(\{\omega\}) = \frac{1}{36}$ for each $\omega \in \Omega$.

**1.4** $\Omega = \{(i,a) : i = 1, \dots, 6, \ a = H \text{ or } T\}$, $P(\{\omega\}) = \frac{1}{12}$ for each $\omega \in \Omega$.

**1.5** We take $\Omega = \{Y, N\}$ if we are only interested in a Yes or No answer to the question of whether the mug will break or not.

The number of pieces after landing can be represented by $\Omega = \{1, 2, \dots\}$.

We can also take $\Omega = \{U, D, R, L\}$ for the landing positions: up, down, handle to the right, or to the left, respectively.

A more sophisticated choice is $\Omega = \{U, D\} \cup [0, 2\pi]$ for a mug with no handle, but with a dot on its side. It may land up, down, or sideways and then roll, in which case we find the angle between the floor and the straight line containing the radius determined by the dot.

In each case, the choice of a probability measure depends on the physical properties of the mug and the floor.

**1.6** There are two possibilities:

(i) We draw two cards at once; $\Omega$ consists of all two-element subsets $\{c_1, c_2\}$ of $C$.

(ii) We draw the second card after having returned the first one to the pack. Then, $\Omega$ consists of all ordered pairs $(c_1, c_2)$ of cards $c_1, c_2 \in C$.

Unless the cards are drawn by a magician, we can take the measure $P(\{\omega\}) = \frac{1}{n}$, where $n$ denotes the number of elements of $\Omega$.

**1.7** $\Omega$ consists of all permutations of the set $\{1, \dots, 6\}$. Each outcome is equally likely, so we take the uniform probability $P(\{\omega\}) = \frac{1}{6!}$ for each $\omega \in \Omega$.

**1.8** The outcomes are all three-, four-, and five-element sequences of letters $W$ or $L$, each containing exactly three letters of one kind (20 possibilities). The uniform probability does not seem appropriate. Everything depends on the probability of winning a single game. Even if we assume that it does not change from game to game and is known, it is not obvious what the measure should be. We shall learn later how to deal with cases of this kind.

**1.9** $\Omega = \{0, 1, 2, \dots\}$, the number of tails before the first head appears. $P(\{0\}) = \frac{1}{2}$, $P(\{1\}) = \frac{1}{4}$, $P(\{2\}) = \frac{1}{8}$, and so on.

We could also add the outcome corresponding to the fact that the head never appears. This is theoretically possible, so put $\Omega' = \Omega \cup \{\infty\}$. The above assignment, $P(\{k\}) = \frac{1}{2^{k+1}}$, remains valid. Because $P(\Omega) = 1$, this leaves us with the only possible choice, $P(\{\infty\}) = 0$. This is an example of an event which theoretically cannot be excluded, but happens with probability 0. We have used the fact that $P(\Omega) = \frac{1}{2} + \frac{1}{4} + \frac{1}{8} + \cdots + \frac{1}{2^n} + \cdots = 1$. In Chapter 6, it will be explained in detail under what circumstances to use such countable sums of probabilities.

**1.10** An idealistic (but mathematically convenient) approach is to assume that the temperature is a real number, so $\Omega = (-\infty, +\infty)$ with probability measure given by a density supplied by the meteorological office (see Example 1.2 for the meaning of density). This density depends on the country in which you live. If the accuracy of meteorological data is to within 1°, we may prefer $\Omega = \{\cdots, -2, -1, 0, +1, +2, \cdots\} = \mathbb{Z}$.

**1.11** If, for example, the line is serviced by six buses and each takes 60 minutes for a complete round, then $\Omega = [0, 10]$ with uniform density (see Example 1.2) seems reasonable. If the bus has a timetable, a density with a peak at each appropriate time may be better.

**1.12** Royal Mail-preferred solution: $\Omega = \{1\}$ if you use a first-class stamp and it is not snowing.

**1.13** We may take $\Omega = \{0, 1, \ldots, n\}$, with $n$ being the capacity of the page. We would not recommend the uniform measure here. $\Omega = \mathbb{N}$ may be more convenient to avoid the dependence on $n$. One may expect the probabilities to decrease as $k$ increases, where $k$ is the number of misprints. Problems of this kind will be examined later.

**1.14** It may be convenient to take $\Omega = \{0, 1, 2, \ldots\}$ with a special measure such that $P(\{n\}) = e^{\lambda} \frac{\lambda^n}{n!}$ for each $n = 0, 1, 2, \ldots$ . This is called the *Poisson measure* with parameter $\lambda > 0$; cf. Problem 8.45.

**1.15** If it is a watch with a simple digital display, we can take

$\Omega_{\text{hours}} = \{1, \ldots, 12\}$,
$\Omega_{\text{minutes}} = \{0, \ldots, 59\}$,
$\Omega_{\text{seconds}} = \{0, \ldots, 59\}$,

or, combining the above,

$\Omega = \{(h, m, s) : h \in \Omega_{\text{hours}}, m \in \Omega_{\text{minutes}}, s \in \Omega_{\text{seconds}}\}$.

Some hours are more likely than others (daytime, evening), but all values of $m$ and $s$ seem to be equally probable.

For an analog display, we can take the same set $\Omega = [0, 2\pi)$ for hours, minutes, and seconds with different densities (uniform for minutes and seconds).

# 2

# Classical Probability Spaces

## 2.1 Theory and Problems

Here, we consider one of the simplest models of a random experiment with $m$ possible outcomes. We assume (believe) that these outcomes are equally likely, so the probability of an event consisting of $n$ outcomes is simply $\frac{n}{m}$. In the framework of general probability spaces, this means that we assume that $\Omega$ is a finite set (with $m$ elements), $\Sigma = 2^{\Omega}$, and $P(\{\omega\}) = \frac{1}{m}$ for each $\omega \in \Omega$; hence,

$$P(A) = \frac{\#A}{\#\Omega}.$$

We shall call such probability spaces *classical* and refer to $P$ as the *uniform probability measure.*

It is useful to know some formulas for computing the numbers of elements of various finite sets.

**2.1** What is the number of subsets of an $n$-element set?

**2.2** In how many ways can you order an $n$-element set?

**2.3** What is the number of all $k$-element subsets of an $n$-element set?

**2.4** What is the number of all one-to-one mappings from an $n$-element set to an $m$-element set?

**2.5** What is the number of all mappings from an $n$-element set to an $m$-element set?

We introduce some notation:

$$\binom{n}{k} = \frac{n!}{k!(n-k)!} = \frac{n \times (n-1) \times \cdots \times (n-k+1)}{1 \times 2 \times 3 \times \cdots \times k}.$$

This gives, as we can see in Solution 2.3, the number of $k$-element subsets of an $n$-element set and is called the *binomial coefficient.*

**2.6** Use the above problems to justify the following particular case of Newton's binomial formula

$$2^n = \sum_{k=0}^{n} \binom{n}{k}.$$

**2.7** Prove the Van der Monde formula

$$\binom{m+n}{k} = \sum_{i=0}^{k} \binom{m}{i} \binom{n}{k-i}.$$

Now some typical exercises on finding probabilities of events. The scheme is as follows. First, we choose the set $\Omega$ and find the appropriate subset $A \subset \Omega$. Here lies the main difficulty because there are many possibilities and we must be careful to be consistent. (In the solution to the next problem, we provide three different approaches, all leading to the same result.) At the end, we count the elements of $\Omega$ and $A$ and compute the probability of $A$.

**2.8** Ten people are randomly seated at a round table. What is the probability that a particular couple will sit next to each other?

**2.9** If boys and girls are born equally likely, what is the probability that in a family with three children, exactly one is a girl.

**2.10** Two dice are thrown. What is the probability that the total number of dots is

a) equal to 7,
b) equal to 3,
c) greater than 5,
d) an even number.

**2.11** In a lottery, 6 numbers are drawn out of 49. Find the probability that

a) 1, 2, 3, 4, 5, 6 are drawn,
b) 4, 23, 24, 35, 40, 45 are drawn,
c) 44 is one of the numbers drawn.

**2.12** What is the probability that among 25 people, at least 2 have their birthday on the same day of the year.

**2.13** Among $t = 60$ lottery tickets, $w = 20$ win prizes. We buy $b = 6$. What is the probability that $g = 2$ will be winning? Generalize this to arbitrary numbers $t, w, b, g$.

∘ **2.14** In a series of 1000 light bulbs, 2% are defective. What is the probability that among 20 bulbs bought, there are 2 faulty ones?

∘ **2.15** From a bridge deck of 52 cards, we draw 13. What is the probability that we have 5 spades in our hand?

**2.16** A bridge deck of 52 cards is dealt among the players. Suppose that I have 4 spades and the opponents have shown (by bidding) that they have 8 hearts. What is the probability that my partner has at least 3 spades?

*For bridge players.* This problem has practical consequences. Suppose that as South you have

♠ AJ84 ♡ 32 ♢ 54 ♣ KQ952

and the bidding is

| N | E | S | W |
|---|---|---|---|
| | 1♡ | pass | 2♡ |
| pass | pass | ? | |

The risk of takeout double is 3♢ response. The best contract may be 2♠, and the estimation of the probability of a complete misfit (partner holding less than three spades) is important.

**2.17** We draw 5 cards out of a deck of 24. What is the probability that we have three of one kind.

**2.18** In the game of poker played with the 24-card deck you get AAAKJ.
a) What is the probability of getting one Ace if you discard KJ?
b) What is the probability of getting a King or an Ace if you discard J?

∘ **2.19** In the game of poker played with the 24-card deck you get AAKJ9.
a) What is the probability of getting one Ace or two Aces if you discard KJ9?
b) What is the probability of getting a King or an Ace if you discard J?

**2.20** A monkey hits a computer keyboard three times at random. What is the chance of getting a three-letter word with a consonant followed by two vowels? The word does not have to make sense. For simplicity, assume that there are 100 keys.

**2.21** From a pack of 52 cards, we draw one-by-one. What is the probability that an Ace will appear at the fifth turn?

**2.22** How likely is it that the word ABRACADABRA will show if the letters A, A, A, A, A, B, B, C, D, R, R are shuffled randomly?

## 2.2 Hints

**2.1** It suffices to consider $n$-element sets of the form $\{1, 2, \dots, n\}$. Try 1, 2, or 3-element sets. The set $\{1\}$ has 2 subsets $\emptyset$ and $\{1\}$, the set $\{1, 2\}$ has 4 subsets, and $\{1, 2, 3\}$ has 8 subsets. If you add one new element to a finite set, how many new subsets (i.e., those containing the new element) will arise? Now try to guess the general formula.

**2.2** A *permutation* of a set $A$ is a one-to-one mapping $f : A \to A$.

If $A = \{1, 2, \dots, n\}$, then the sequence $f(1), f(2), \dots, f(n)$ contains all the elements of $A$. So one can say that a permutation of a finite set is the order in which the elements are arranged.

Try to figure out the general formula by finding all permutations of 1-, 2-, 3-element sets. How many new permutations will arise if you add one new element to a finite set?

**2.3** Consider an example. For $A = \{1, 2, 3, 4, 5\}$, count the number of 3-element subsets of $A$. The first element of a subset can be chosen in 5 ways, the second in 4, and the third in 3 ways, giving in total $5 \times 4 \times 3 = 60$. But this is incorrect! Each subset has been counted 6 times, so we have to divide 60 by 6. To see this, consider, for example, $\{1, 3, 4\}$. The elements of this set could have been picked in any order. So the following sets emerge from this method of counting: $\{1, 3, 4\}$, $\{1, 4, 3\}$, $\{3, 1, 4\}$, $\{3, 4, 1\}$, $\{4, 1, 3\}$, $\{4, 3, 1\}$. But they are all identical!

**2.4** Suppose we are counting the number of one-to-one mappings $f$ from the set $\{1, 2, 3\}$ to $\{1, 2, 3, 4, 5\}$. First, $f(1)$ can be chosen in 5 ways, then $f(2)$ in 4 ways, and, finally, $f(3)$ in 3 ways. The total number is $5 \times 4 \times 3$.

**2.5** Suppose we are counting the number of mappings $f$ from $\{1, 2, 3\}$ to $\{1, 2, 3, 4, 5\}$. We can choose $f(1)$ in 5 ways, then $f(2)$ in 5 ways, and, finally, $f(3)$ also in 5 ways (the values may coincide). The total number is $5 \times 5 \times 5 = 5^3$.

**2.6** Use Problems 2.1 and 2.3.

**2.7** Imagine choosing $k$ balls, among which are $m$ white balls and $n$ black balls. You could do that, for example, by choosing $i$ white balls and $k-i$ black balls for some $i$ between 0 and $k$. Counting the number of choices in this way, you can obtain the right-hand side of Van der Monde's formula. To obtain the left-hand side, refer to Problem 2.3.

**2.8** How in practice can we seat 10 people randomly at the table?

1. We number the seats from 1 to 10, prepare a deck of 10 cards with the same numbers on them, shuffle, then deal among the people. The number of outcomes is the same as the number of all permutations of 10 elements. Count the "favorable" outcomes that the two people are seated next to each other.

2. Prepare the cards and the seats as before. Now deal 2 cards to the couple. This can be done in the same number of ways as the number of 2-element subsets of a 10-element set. How many sets are "favorable"?

**2.9** Consider the tree in Figure 2.1.

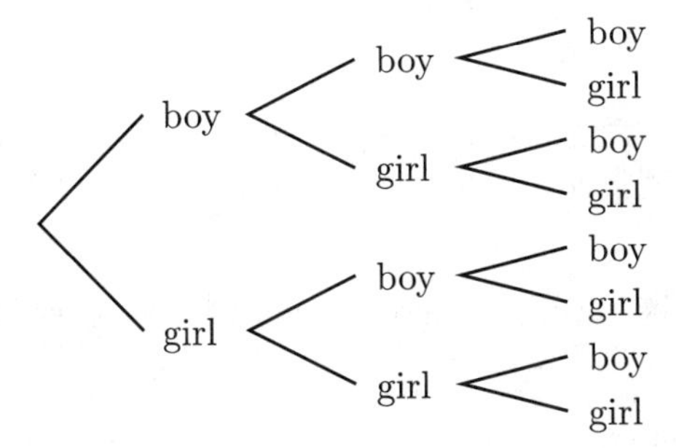


FIGURE 2.1. Three children

All branches are equally likely. How many "favorable" ones are there?

**2.10** All possible outcomes are shown in Figure 2.2.

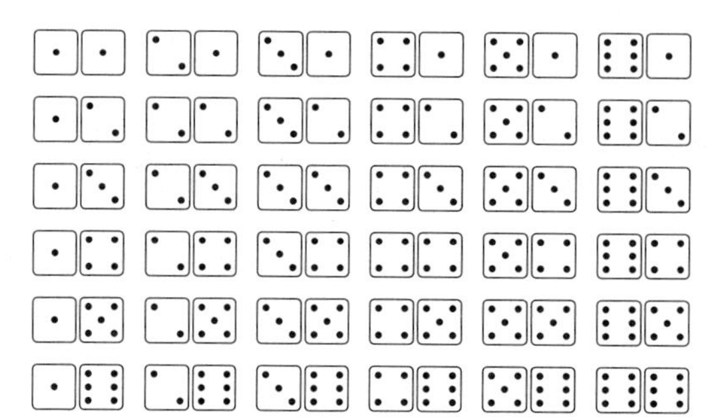

FIGURE 2.2. Two dice

All possible sums are in the table:

| | 1 | 2 | 3 | 4 | 5 | 6 |
|---|---|---|---|---|---|---|
| 1 | 2 | 3 | 4 | 5 | 6 | 7 |
| 2 | 3 | 4 | 5 | 6 | 7 | 8 |
| 3 | 4 | 5 | 6 | 7 | 8 | 9 |
| 4 | 5 | 6 | 7 | 8 | 9 | 10 |
| 5 | 6 | 7 | 8 | 9 | 10 | 11 |
| 6 | 7 | 8 | 9 | 10 | 11 | 12 |

Some sums appear more frequently than others. It would be ill-advised to take $\Omega = \{2, 3, \ldots, 11, 12\}$ with uniform probability measure.

**2.11** The order is irrelevant, so the model should be based on selecting 6-element subsets.

**2.12** It is best to consider the opposite first: the probability that all 25 people have their birthdays on different days. Disregard leap years for simplicity.

**2.13** Buying tickets is like selecting a subset. The same can be said about winning and losing tickets.

**2.14** See Problem 2.13, in particular the general formula given in the solution. Here, a bulb is a ticket, a defective bulb is a prize, and buying is the same as drawing at random.

**2.15** See Problem 2.13.

**2.16** The situation is similar to that in Problem 2.15. It is best to begin with computing the probability that the partner has 0, 1, or 2 spades.

**2.17** Drawing 5 cards means choosing a 5-element subset. There are 6 possible triplets: 3 Aces, Kings, Queens, Jacks, Tens, or Nines.

**2.18** You draw 1 or 2 out of the remaining 19 cards, so the sample space consists of 1- or 2-element subsets. You have to count the "favorable" outcomes carefully.

**2.19** See Problem 2.18.

**2.20** The sample space consists of all functions from the set $\{1, 2, 3\}$ to $\{1, 2, \ldots, 100\}$. Count the number of "favorable" ones.

**2.21** Since the order is relevant, we consider functions (sequences) rather than sets.

**2.22** If all the letters had been different, we would have a 1 in 11! chance. But in our case, there are multiple instances of some letters and certain permutations lead to the same word.

## 2.3 Solutions

**2.1** The answer is $2^n$. We shall prove this by induction. We can assume without loss of generality that the $n$-element set is of the form $\{1, \ldots, n\}$.

The case $n = 1$ was discussed in the hint.

Induction hypothesis: Suppose that for some $n = k$, the $k$-element set $\{1, 2, \ldots, k\}$ has $2^k$ subsets. Consider the $(k+1)$-element set $\{1, 2, \ldots, k, k+1\}$. Each subset $A$ of $\{1, 2, \ldots, k, k+1\}$ is either contained in $\{1, 2, \ldots, k\}$ or not depending on whether $k+1$ belongs to $A$ or not. The number of subsets $A$ contained in $\{1, 2, \ldots, k\}$ is $2^k$ by the induction hypothesis. The number of subsets $A$ contained in $\{1, 2, \ldots, k, k+1\}$ but not in $\{1, 2, \ldots, k\}$ is also $2^k$ because the mapping $A \mapsto A \cup \{k+1\}$ defines a one-to-one correspondence between these two kinds of subsets. This means that the total number of subsets of the $(k+1)$-element set $\{1, 2, \ldots, k, k+1\}$ is $2^k + 2^k = 2^{k+1}$, completing the proof.

**2.2** The answer is $n! = 1 \times 2 \times \cdots \times n$.

To see this, let us count all possible permutations of $\{1, 2, \ldots, n\}$ by putting its elements into numbered cells. Number 1 can be placed in $n$ cells, which leaves $n - 1$ free cells in which to put the next number. This means that there are $n \times (n-1)$ ways in which 1 and 2 can be placed in the cells. Next, we have $n - 2$ free cells in which to put 3, which gives $n \times (n-1) \times (n-2)$ ways of placing 1, 2, and 3. After placing all $n$ numbers in this manner, we see that this can be done in $n \times (n-1) \times \cdots \times 2 \times 1 = n!$ ways, as claimed.

**2.3** The answer is

$$\frac{n \times (n-1) \times \cdots \times (n-k+1)}{1 \times 2 \times 3 \times \cdots \times k}.$$

To select a $k$-element subset, choose the elements one-by-one. The first can be chosen in $n$ ways, the second in $n-1$ ways, and so on. The number of all possibilities is the product in the numerator. This method of counting distinguishes between $k$-element sequences ordered in a different way. To obtain the number of $k$-element subsets (where the order of elements does not matter), we have to divide by the number of permutations of a $k$-element set, i.e., by $k! = 1 \times 2 \times 3 \times \cdots \times k$.

**2.4** The answer is $m \times (m-1) \times (m-2) \times \cdots \times (m-n+1)$. Without loss of generality, we can assume that the domain of our mappings is the set $\{1, 2, 3, \ldots, n\}$ and the range is $\{1, 2, 3, \ldots, m\}$. The value $f(1)$ can be selected in $m$ ways, $f(2)$ in $m-1$ ways, and so on. After $n$ steps, we have the total number of possibilities as claimed.

**2.5** The answer is $m^n$. Without loss of generality, we can assume that the domain of our mappings is the set $\{1, 2, 3, \ldots, n\}$ and the range is $\{1, 2, 3, \ldots, m\}$. The value $f(1)$ can be selected in $m$ ways, $f(2)$ also in $m$ ways (it is possible that $f(1) = f(2)$), and so on. After $n$ steps, we get the result.

**2.6** The left-hand side of the equality to be verified is the number of all subsets of an $n$-element set. On the right, we have the sum of the numbers of 0-, 1-, 2-, ..., $n$-element sets. Together, they exhaust all possible subsets.

**2.7** Consider an $(m+n)$-element set $A$. The number of $k$-element subsets in $A$ is $\binom{m+n}{k}$; see Problem 2.3. Now, take an $m$-element set $B$ and an $n$-element set $C$ such that $A = B \cup C$ and consider those $k$-element subsets in $A$ that have $i$ elements in $B$ and $k-i$ elements in $C$. There are $\binom{m}{i}\binom{n}{k-i}$ such subsets, again by Problem 2.3. The sum from $i = 0$ to $k$ gives the number of all $k$-element subsets in $A$, which proves Van der Monde's formula.

**2.8** We give 3 solutions depending on the model for seating 10 persons randomly at a round table.

1. Having numbered the seats from 1 to 10, we shuffle and deal a deck of 10 cards with these numbers. Here, $\Omega$ is the set of all permutations of a 10-element set, so $\#\Omega = 10!$. A permutation is "favorable" when the couple sit next to each other, i.e., draw $(1, 2)$ or $(2, 3)$ or ... or $(9, 10)$ or $(10, 1)$ (the table is round), or the inverted pairs $(2, 1)$ or $(3, 2)$ or ... or $(1, 10)$. This gives 20 possibilities. The

remaining 8 people can be seated in any order, so we multiply by 8!. The required probability is

$$\frac{20 \times 8!}{10!} = \frac{2}{9}.$$

2. We consider the cards drawn by the couple as a set (the order does not matter) and we are not bothered with the rest of the people. So, $\Omega$ is the set of all 2-element subsets of a 10-element set with $\#\Omega = \binom{10}{2}$. There are 10 "favorable" sets: $\{1,2\}$, $\{2,3\}$, ... , $\{10,1\}$, so the probability is

$$\frac{10}{\binom{10}{2}} = \frac{10}{\frac{10\times 9}{1\times 2}} = \frac{2}{9}.$$

3. The neatest solution which comes to mind is this. One partner draws a card. There are 9 left with 2 "favorable" ones, so the probability that the couple will sit together is $\frac{2}{9}$, as before.

**2.9** There are eight possibilities:

BBB, BBG, BGB, BGG, GGG, GGB, GBG, GBB,

out of which three are "favorable." The probability is $\frac{3}{8}$.

**2.10** Counting the "favorable" outcomes in Figure 2.2, we easily get the following answers:

a) $\frac{6}{36} = \frac{1}{6}$, c) $\frac{26}{36} = \frac{13}{18}$,
b) $\frac{2}{36} = \frac{1}{18}$, d) $\frac{18}{36} = \frac{1}{2}$.

**2.11** $\Omega$ consists of all 6-element subsets of the set $\{1, 2, \ldots, 49\}$. Thus, $\#\Omega = \binom{49}{6}$. In cases a) and b), we have one "favorable" outcome, so the probability is

$$\frac{1}{\binom{49}{6}} = \frac{1}{\frac{49\times 48\times 47\times 46\times 45\times 44}{1\times 2\times 3\times 4\times 5\times 6}} = \frac{1}{\frac{10068347520}{720}} = \frac{1}{13983816}.$$

In case c), the number of selections containing a specific number, 44 in this case, is equal to the number of ways the remaining 5 numbers can be chosen out of 48, which is $\binom{48}{5}$. The probability is equal to

$$\frac{\binom{48}{5}}{\binom{49}{6}} = \frac{\frac{48\times 47\times 46\times 45\times 44}{1\times 2\times 3\times 4\times 5}}{\frac{49\times 48\times 47\times 46\times 45\times 44}{1\times 2\times 3\times 4\times 5\times 6}} = \frac{6}{49}.$$

**2.12** The desired probability is equal to $p = 1 - q$, where $q$ is the probability that all 25 people have birthdays on different days.

The sample space is the set of all functions from $\{1, 2, \dots, 25\}$ to $\{1, 2, \dots, 365\}$ with $\#\Omega = 365^{25}$ (we disregard leap years for simplicity). The number of assignments with different birthdays is the same as the number of one-to-one mappings: $365 \times 364 \times \cdots \times 341$. Dividing these two, we have

$$q = \frac{365 \times 364 \times \cdots \times 341}{365^{25}} = \frac{365}{365} \times \frac{364}{365} \times \cdots \times \frac{341}{365} \approx 0.40,$$

so $p \approx 0.60$, which is surprisingly large.

**2.13** The sample space consists of all 6-element subsets of a 60-element set, so $\#\Omega = \binom{60}{6}$. Our 2 winning tickets are 2-element subsets of the set of 20. To each such a selection, there correspond $\binom{40}{4}$ ways in which the remaining, losing, tickets can be selected. It follows that

$$p = \frac{\binom{20}{2} \times \binom{40}{4}}{\binom{60}{6}} \approx 0.35.$$

The general formula has the form

$$p = \frac{\binom{w}{g} \times \binom{t-w}{b-g}}{\binom{t}{b}}.$$

**2.16** The sample space consists of all 13-element subsets of a 31-element set ($52 - 13 - 8$, excluding our hand and the hearts shown), so $\#\Omega = \binom{31}{13}$. The number of hands with 0 spades is $\binom{22}{13}$ (since there are 9 spades among the 31 cards), with 1 spade $9 \times \binom{22}{12}$, and with 2 spades $\binom{9}{2} \times \binom{22}{11}$. So the answer is

$$p = 1 - \frac{\binom{22}{13} + 9 \times \binom{22}{12} + \binom{9}{2} \times \binom{22}{11}}{\binom{31}{13}} \approx 0.85.$$

**2.17** The sample space consists of all 5-element subsets of the 24-element set of cards, so $\#\Omega = \binom{24}{5}$. Three aces can be selected in $\binom{4}{3}$ ways and the remaining 2 cards in $\binom{20}{2}$ ways. Multiply this by 6 to get

$$p = \frac{6 \times \binom{4}{3} \times \binom{20}{2}}{\binom{25}{5}} \approx 0.086.$$

**2.18** a) We have $\#\Omega = \binom{19}{2} = 171$ and there are 18 "favorable" outcomes: the remaining Ace with any of the other 18 cards. So $p = \frac{18}{171} \approx 0.1052$.

b) We have $\#\Omega = \binom{19}{1} = 19$ and there are 4 "favorable" outcomes: three Kings and the remaining Ace, so $p = \frac{4}{19} \approx 0.2105$.

**2.20** Taking $\Omega$ to be the set of all functions from $\{1, 2, 3\}$ to $\{1, 2, \dots, 100\}$, we have $\#\Omega = 100^3$. There are 20 consonants and 6 vowels. The number of "favorable" outcomes is $20 \times 6 \times 6$ and the probability is

$$p = \frac{20 \times 6 \times 6}{100^3} = 0.00072.$$

**2.21** Consider 5-element sequences of cards, i.e., one-to-one mappings from $\{1, 2, 3, 4, 5\}$ to $\{1, 2, \dots, 52\}$ with $\#\Omega = 52 \times 51 \times 50 \times 49 \times 48$. To represent the event in question, we count all sequences of 4 cards out of 48 and multiply by 4, as there are 4 aces in the deck. Hence,

$$p = \frac{48 \times 47 \times 46 \times 45 \times 4}{52 \times 51 \times 50 \times 49 \times 48} \approx 0.05989.$$

**2.22** Suppose that we label each letter with a number from 1 to 11. The sample space is the set of permutations of 11 numbers, so $\#\Omega = 11!$. We count the permutations giving the required word: The letters A can be rearranged in 5! ways, and B and R in 2! ways each, so we obtain $5! \times 2! \times 2!$ "favorable" outcomes and

$$p = \frac{5! \times 2! \times 2!}{11!} = \frac{1}{83160}.$$

# 3

# Fields

## 3.1 Theory and Problems

In this chapter, we shall formalize the concept of events discussed in Chapter 1. We want events to be subsets of the set $\Omega$ that belong to the domain of a probability measure $P$, which means that if $A$ is an event, then it makes sense to compute $P(A)$. It is natural to ask questions like this: What is the probability of the union of two events (one or the other happens)? What is the probability of the intersection of two events (both happen)? What is the probability of the complement of an event (the event does not happen)? Therefore, we require that the family of sets in question should be closed under these operations on sets. This gives rise to the notion of a field of sets (events).

▶ **Definition 3.1** A family $\mathcal{F}$ of subsets of a non-empty set $\Omega$ is called a *field* on $\Omega$ if

(a) $\Omega \in \mathcal{F}$;

(b) if $A \in \mathcal{F}$, then $\Omega \setminus A \in \mathcal{F}$;

(c) if $A, B \in \mathcal{F}$, then $A \cup B \in \mathcal{F}$.

The elements of a field are called *events*.

The conditions of this definition can be read as follows:

(a) the set of all outcomes is an event;

(b) the complement of an event is an event;

(c) the union of two events is an event.

We have imposed no condition involving the intersection of sets. A condition like this can be deduced from the others; see Problem 3.11.

We begin with some examples. Show that

**3.1** $\{\emptyset, \Omega\}$ is a field;

**3.2** The family of all subsets of a finite set $\Omega$ is a field;

**3.3** If $A \subset \Omega$, then $\{\emptyset, \Omega, A, \Omega \setminus A\}$ is a field.

**3.4** Let $\Omega = \{1, 2, 3, 4\}$. Is any of the following families of sets a field:
$\mathcal{F}_1 = \{\emptyset, \{1, 2\}, \{3, 4\}\}$,
$\mathcal{F}_2 = \{\emptyset, \Omega, \{1\}, \{2, 3, 4\}, \{1, 2\}, \{3, 4\}\}$,
$\mathcal{F}_3 = \{\emptyset, \Omega, \{1\}, \{2\}, \{1, 2\}, \{3, 4\}, \{2, 3, 4\}, \{1, 3, 4\}\}$?

**3.5** Let $\Omega = (0, 1)$. Is any of the following families of sets a field:
$\mathcal{F}_1 = \{\emptyset, (0, 1), (0, \frac{1}{2}), (\frac{1}{2}, 1)\}$,
$\mathcal{F}_2 = \{\emptyset, (0, 1), (0, \frac{1}{2}), [\frac{1}{2}, 1), (0, \frac{2}{3}], (\frac{2}{3}, 1)\}$,
$\mathcal{F}_3 = \{\emptyset, (0, 1), (0, \frac{2}{3}), [\frac{2}{3}, 1)\}$?

**3.6** Let $\Omega = [0, 1]$. Adding as few sets as possible, complete the family of sets $\{\emptyset, [0, \frac{1}{2}), \{1\}\}$ to obtain a field.

**3.7** Let $\Omega = \{1, 2, 3\}$. Complete $\{\{2\}, \{3\}\}$ to obtain a field. Add as few sets as possible.

The following problems develop some basic properties of fields. We shall see that fields are closed under all finite set-theoretic operations. If $\mathcal{F}$ is a field, show that

▶ **3.8** $\emptyset \in \mathcal{F}$;

**3.9** If $A, B, C \in \mathcal{F}$, then $A \cup B \cup C \in \mathcal{F}$;

**3.10** For each natural number $n$, if $A_1, \ldots, A_n \in \mathcal{F}$, then $A_1 \cup \cdots \cup A_n \in \mathcal{F}$;

▶ **3.11** If $A, B \in \mathcal{F}$, then $A \cap B \in \mathcal{F}$;

**3.12** For each natural number $n$, if $A_1, \ldots, A_n \in \mathcal{F}$, then $A_1 \cap \cdots \cap A_n \in \mathcal{F}$;

▶ **3.13** If $A, B \in \mathcal{F}$, then $A \setminus B \in \mathcal{F}$;

**3.14** If $A, B \in \mathcal{F}$, then their symmetric difference $A \triangle B$ also belongs to $\mathcal{F}$.

**3.15** Prove that if $\mathcal{F}_1$ and $\mathcal{F}_2$ are fields of subsets of $\Omega$, then $\mathcal{F}_1 \cap \mathcal{F}_2$ is also a field.

**3.16** Find two fields such that their union is not a field.

So far we have considered examples of finite fields. There are also fields consisting of infinitely many events.

**3.17** Is $\mathcal{F} = \{A \subset \Omega : A \text{ is a finite set}\}$ always a field?

**3.18** Let $I_0 = \emptyset$ and $I_n = \{1, \ldots, n\}$ for $n \geqslant 1$. Is $\mathcal{F} = \{I_n\}_{n=0}^{\infty} \cup \{\mathbb{N} \setminus I_n\}_{n=0}^{\infty}$ a field?

**3.19** Let $\Omega$ be a (possibly infinite) set. Show that the family $\mathcal{F}$ consisting of all finite subsets of $\Omega$ and all subsets of $\Omega$ having a finite complement is a field.

$*$ **3.20** We say that $A \subset \mathbb{Z}$ is a periodic set if there is a positive integer $i$ and a set $I \subset \{1, \ldots, i\}$ such that

$$A = \bigcup_{l=-\infty}^{+\infty} (I + li), \tag{3.1}$$

where $I + li = \{n + li : n \in I\}$.

Show that

$$\mathcal{F} = \{A \subset \mathbb{Z} : A \text{ is a periodic set}\}$$

is a field on $\mathbb{Z}$.

The smallest (with respect to inclusion) non-empty events belonging to $\mathcal{F}$ are called *atoms*. This notion is particularly useful when dealing with finite fields.

**3.21** What are the atoms in the fields
$\mathcal{F}_1 = \{\emptyset, \{1\}, \{2\}, \{1,2\}, \{3,4\}, \{2,3,4\}, \{1,3,4\}, \{1,2,3,4\}\}$,
$\mathcal{F}_2 = \{\emptyset, (0,1), (0, \frac{2}{3}), [\frac{2}{3}, 1)\}$?

**3.22** What are the atoms in the field of all subsets of a finite set $\Omega$?

**3.23** Show that different atoms must be disjoint.

**3.24** Show that if $\mathcal{F}$ is a finite field, then every non-empty event $A \in \mathcal{F}$ contains an atom.

**3.25** Show that if $\mathcal{F}$ is a finite field, then each event $A$ in $\mathcal{F}$ is the union of finitely many atoms.

Show that this is a unique representation; that is, the set of atoms $\{A_1, \ldots, A_k\}$ such that $A = A_1 \cup \cdots \cup A_k$ is unique for each event $A$ in $\mathcal{F}$.

We are now in a position to prove a remarkable result that a finite field can only have a very special number of elements.

**3.26** Show that every finite field has $2^n$ elements for some $n \in \mathbb{N}$.

Problems 3.4 and 3.5 can now be solved without having to verify the conditions of Definition 3.1 for all possible combinations of events. Instead, use what you have just learned about atoms. Here is one more problem of this kind.

**3.27** Which of the following families of sets is a field:
$\mathcal{F}_1 = \{\emptyset, \{a\}, \{b\}, \{a, c\}, \{a, b, c\}\}$,
$\mathcal{F}_2 = \{\emptyset, [0, 1], (\frac{1}{4}, 1], [0, \frac{1}{4}], [0, \frac{1}{4}), \{\frac{1}{4}\}, [\frac{1}{4}, 1], [0, 1] \setminus \{\frac{1}{4}\}\}$,
$\mathcal{F}_3 = \{\emptyset, \{1\}, \{2\}, \{3\}, \{1, 2\}, \{2, 3\}, \{2, 3, 4\}, \{1, 2, 3\}\}$?

We close this chapter with an example of a field which has no atoms.

$*$ **3.28** Show that the field of periodic sets in Problem 3.20 has no atoms.

## 3.2 Hints

**3.1** It is only conditions (b) and (c) of Definition 3.1 whose verification may cause any problems. You need to consider two cases $A = \emptyset$ and $A = \Omega$ to verify condition (b). To verify (c), consider three cases:
$A = \emptyset$, $B = \emptyset$;
$A = \emptyset$, $B = \Omega$;
$A = \Omega$, $B = \Omega$.

**3.2** Here, $\mathcal{F}$ consists of all subsets of $\Omega$. So the condition $A \in \mathcal{F}$ reads: $A$ is a subset of $\Omega$.

**3.3** See Hint 3.1. However, more cases need to be considered.

**3.4** All conditions can be checked by analyzing finitely many cases.

**3.5** See Hints 3.3 and 3.4.

**3.6** If you add some sets to fulfill (a), then (b), and, finally, (c) of Definition 3.1, remember that anything added at stage (c) may have destroyed (b), so it may be necessary to rerun the procedure.

**3.7** See Hint 3.6.

**3.8** Use (a) and (b) of Definition 3.1.

**3.9** We know that the union of two sets is in $\mathcal{F}$. Use this property twice.

**3.10** Use (c) of Definition 3.1 repeatedly or apply induction on $n$.

**3.11** How can we express the intersection of two sets if we are allowed to use complements and unions only?

**3.12** Use either Problem 3.11 and induction on $n$, or Problem 3.10 and de Morgan's law.

**3.13** How can we express the difference of two sets using unions and complements? (If you have done Problem 3.11, you may also use intersections.)

**3.14** Let us recall the definition of symmetric difference: $A \triangle B = (A \setminus B) \cup (B \setminus A)$. Now use Problem 3.13 and condition (c) of Definition 3.1.

**3.15** For example, we show that $\mathcal{F}_1 \cap \mathcal{F}_2$ has property (b) of Definition 3.1. Suppose that $A \in \mathcal{F}_1 \cap \mathcal{F}_2$. Hence, $A \in \mathcal{F}_1$ and $A \in \mathcal{F}_2$. Condition (b) of Definition 3.1 is valid for both $\mathcal{F}_1$ and $\mathcal{F}_2$, so it can be used to show that $\Omega \setminus A \in \mathcal{F}_1 \cap \mathcal{F}_2$.

**3.16** Try two different 4-element fields. The example will be elegant if you keep $\Omega$ as small as possible.

**3.17** What happens if $\Omega$ is an infinite set?

**3.18** If you take $I_m$ and $\mathbb{N} \setminus I_n$ for some $m$ and $n$, is their union always in $\mathcal{F}$?

**3.19** Conditions (a) and (b) of Definition 3.1 follow immediately. Condition (c) involves two sets in $\mathcal{F}$. Consider two cases: when both sets are finite and when at least one of them has a finite complement.

**3.20** You have to show that $\emptyset$, $\mathbb{Z}$, $\mathbb{Z} \setminus A$, and $A \cup B$ are periodic sets if $A$ and $B$ are periodic. Try to represent each of the four sets as a union of the form (3.1) for a suitable $i$ and $I \subset \{1, \ldots, i\}$. This is not hard in the case of $\emptyset$ and $\mathbb{Z}$. In the case of $\mathbb{Z} \setminus A$ and $A \cup B$, try to find the corresponding $i$ and $I$ in terms of those of $A$ and $B$.

You may think of $i$ in (3.1) as the period of a periodic set. If $i$ is the period, what is the role of $I$ in (3.1)? (If you prefer to work with periodic functions rather than sets, consider the indicator function of a periodic set.) Refer to Figure 3.1 for an example of a periodic set.

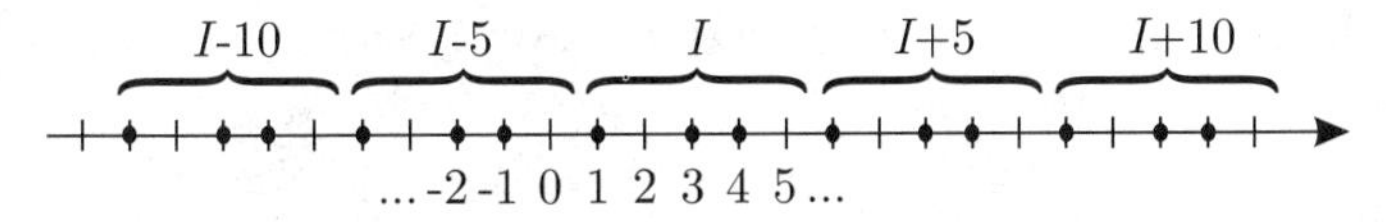


FIGURE 3.1. Example of a periodic set with $i = 5$ and $I = \{1, 3, 4\}$

If $i$ is the period of $A$, what is the period of $\mathbb{Z} \setminus A$? If $i$ is the period of $A$ and $j$ the period of $B$, what is the period of $A \cup B$? Once you have decided what the period of $\mathbb{Z} \setminus A$ and $A \cup B$ should be, it becomes easier to represent these sets in the form (3.1).

**3.21** Are $\mathcal{F}_1$ and $\mathcal{F}_2$ fields in the first place? What are the non-empty sets in $\mathcal{F}_i$ that do not contain any other sets in $\mathcal{F}_i$?

**3.22** What are the smallest (with respect to inclusion) non-empty subsets in $\Omega$? How many elements do they have?

**3.23** Atoms are the smallest events with respect to inclusion. What does this imply about the intersection of two atoms, which is contained in both of them?

**3.24** Take any $\omega \in A$ and consider the intersection of all events containing $\omega$. Is this intersection an event? Is it contained in $A$? Is it an atom?

**3.25** What is the union of all atoms contained in $A$? To show uniqueness, bear in mind that atoms are non-empty and pairwise disjoint.

**3.26** Events in a finite field are built from atoms. How many different events can you build from $n$ atoms?

**3.27** Count the elements of each family of sets. You will be able to discard some families at this stage. Then, identify and count the atoms. Can every set in the family under consideration be built from these atoms?

**3.28** For every non-empty periodic set $A$, you need to find another non-empty periodic set $B \neq A$ such that $B \subset A$. If $A$ is given by (3.1), you may try to remove one or more points from the corresponding set $I$. But what if $I$ is a one-element set? How about doubling the period and removing a point from every other copy of $I$?

## 3.3 Solutions

**3.1** We verify the conditions of Definition 3.1. The first is obvious by inspection. The second reduces to two cases: $A = \emptyset$ or $A = \Omega$. But $\Omega \setminus \emptyset = \Omega$ and $\Omega \setminus \Omega = \emptyset$, so in both cases, the complement is in $\mathcal{F}$. Finally, to verify the third condition, we consider the three cases listed in the hint. Because $\emptyset \cup \emptyset = \emptyset$, $\emptyset \cup \Omega = \Omega$, and $\Omega \cup \Omega = \Omega$, in each case the result is in $\mathcal{F}$.

**3.2** Here, $\mathcal{F}$ contains all subsets of $\Omega$; hence, it also contains $\Omega$, $\Omega \setminus A$, and $A \cup B$, no matter what $A$ and $B$ are.

**3.3** Condition (a) of Definition 3.1 can be verified by inspection. For (b), we find the complement of each set: $\Omega \setminus \emptyset = \Omega$, $\Omega \setminus \Omega = \emptyset$, $\Omega \setminus A$, and $(\Omega \setminus (\Omega \setminus A)) = A$, so (b) is satisfied.

For (c), take the union of any two sets in $\{\emptyset, \Omega, A, \Omega \setminus A\}$. If one of the sets is $\Omega$, then the union is also $\Omega$. If one of the sets is $\emptyset$, then the union is the other set. If both sets are identical, then their union is just the set: $A \cup A = A$ and $(\Omega \setminus A) \cup (\Omega \setminus A) = \Omega \setminus A$. There is only one case left: $A \cup (\Omega \setminus A) = \Omega$. Condition (c) is satisfied in each case.

**3.4** $\mathcal{F}_1$ is not a field because $\Omega \notin \mathcal{F}_1$.
$\mathcal{F}_2$ does not satisfy (c), since $\{1\} \cup \{3,4\} \notin \mathcal{F}_2$.
$\mathcal{F}_3$ is a field. All conditions are easy to verify.

**3.5** $\mathcal{F}_1$ is not a field because $\Omega \setminus (0, \frac{1}{2}) = [\frac{1}{2}, 1) \notin \mathcal{F}_1$.
$\mathcal{F}_2$ is not a field because $(0, \frac{1}{2}) \cup (\frac{2}{3}, 1) \notin \mathcal{F}_2$.
$\mathcal{F}_3$ is a field; see Solution 3.3.

**3.6** $\mathcal{F} = \{\emptyset, [0,1], [0, \frac{1}{2}), [\frac{1}{2}, 1], \{1\}, [0,1), [0, \frac{1}{2}) \cup \{1\}, [\frac{1}{2}, 1)\}$.

**3.7** $\mathcal{F} = \{\emptyset, \{1,2,3\}, \{2\}, \{1,3\}, \{3\}, \{1,2\}, \{2,3\}, \{1\}\}$.

**3.8** Using (b) from Definition 3.1 for $A = \Omega$, which by (a) is in $\mathcal{F}$, we get $\Omega \setminus \Omega = \emptyset \in \mathcal{F}$.

**3.9** First, $A \cup B \in \mathcal{F}$ by condition (c) of Definition 3.1. Next, since the two sets $A \cup B$ and $C$ are in $\mathcal{F}$, their union $(A \cup B) \cup C = A \cup B \cup C$ is in $\mathcal{F}$ by (c), as required.

**3.10** We shall use induction on $n$. For $n = 1$, the claim is obvious.

Induction hypothesis: Suppose that the claim holds for some $n = k$:

$$\text{If } A_1, \ldots, A_k \in \mathcal{F}, \text{ then } A_1 \cup \cdots \cup A_k \in \mathcal{F}.$$

Take $k+1$ sets $A_1, \ldots, A_k, A_{k+1} \in \mathcal{F}$. By the induction hypothesis, the union of the first $k$ sets is in $\mathcal{F}$; that is, $A_1 \cup \ldots \cup A_k \in \mathcal{F}$. We also know that $A_{k+1} \in \mathcal{F}$. Using (c) of Definition 3.1 with $A = A_1 \cup \cdots \cup A_k$ and $B = A_{k+1}$, we get $A \cup B = A_1 \cup \cdots \cup A_{k+1} \in \mathcal{F}$, which proves the claim for $n = k+1$.

**3.11** $A \cap B = \Omega \setminus [(\Omega \setminus A) \cup (\Omega \setminus B)]$, $\Omega$ is in $\mathcal{F}$ by (a) of Definition 3.1, $\Omega \setminus A$ and $\Omega \setminus B$ are in $\mathcal{F}$ by (b), their union is in $\mathcal{F}$ by (c), and its complement is in $\mathcal{F}$ again by (b).

**3.12** If $A_1, \dots, A_n \in \mathcal{F}$, then $\Omega \setminus A_1, \dots, \Omega \setminus A_n \in \mathcal{F}$, so $(\Omega \setminus A_1) \cup \dots \cup (\Omega \setminus A_n) \in \mathcal{F}$ by Problem 3.10. According to de Morgan's law,

$$A_1 \cap \dots \cap A_n = \Omega \setminus ((\Omega \setminus A_1) \cup \dots \cup (\Omega \setminus A_n)).$$

It follows that $A_1 \cap \dots \cap A_n \in \mathcal{F}$, as required.

**3.13** We can write $A \setminus B = A \cap (\Omega \setminus B)$. Then, $\Omega \setminus B \in \mathcal{F}$ by (b) of Definition 3.1 and $A \cap (\Omega \setminus B) \in \mathcal{F}$ by Problem 3.11.

**3.14** $A \setminus B$ and $B \setminus A$ are in $\mathcal{F}$ by Problem 3.13. It follows that $A \triangle B = (A \setminus B) \cup (B \setminus A)$ belongs to $\mathcal{F}$ by condition (c) of Definition 3.1.

**3.15** First, $\Omega \in \mathcal{F}_1$ and $\Omega \in \mathcal{F}_2$, so $\Omega \in \mathcal{F}_1 \cap \mathcal{F}_2$; that is, condition (a) of Definition 3.1 holds for $\mathcal{F}_1 \cap \mathcal{F}_2$.

Let $A \in \mathcal{F}_1 \cap \mathcal{F}_2$. Then, $A \in \mathcal{F}_1$ and $A \in \mathcal{F}_2$; hence, $\Omega \setminus A \in \mathcal{F}_1$ and $\Omega \setminus A \in \mathcal{F}_2$ because (b) is satisfied for $\mathcal{F}_1$ and $\mathcal{F}_2$. Thus, $\Omega \setminus A \in \mathcal{F}_1 \cap \mathcal{F}_2$, which means that condition (b) is satisfied for $\mathcal{F}_1 \cap \mathcal{F}_2$.

Let $A, B \in \mathcal{F}_1 \cap \mathcal{F}_2$. Then, $A, B \in \mathcal{F}_1$ and $A, B \in \mathcal{F}_2$; hence, $A \cup B \in \mathcal{F}_1$ and $A \cup B \in \mathcal{F}_2$, because $\mathcal{F}_1$ and $\mathcal{F}_2$ both satisfy (c). Thus, $A \cup B \in \mathcal{F}_1 \cap \mathcal{F}_2$, so (c) is valid for $\mathcal{F}_1 \cap \mathcal{F}_2$.

**3.16** Let $\Omega = \{1, 2, 3\}$ and let us take $\mathcal{F}_1 = \{\emptyset, \Omega, \{1\}, \{2, 3\}\}$ and $\mathcal{F}_2 = \{\emptyset, \Omega, \{2\}, \{1, 3\}\}$. Then, $\mathcal{F}_1 \cup \mathcal{F}_2 = \{\emptyset, \Omega, \{1\}, \{2\}, \{2, 3\}, \{1, 3\}\}$, which is not a field, since $\{1\} \cup \{2\} = \{1, 2\}$ is not in $\mathcal{F}_1 \cup \mathcal{F}_2$.

**3.17** If $\Omega$ is an infinite set, then $\Omega \notin \mathcal{F}$, so condition (a) of Definition 3.1 does not hold.

**3.18** Condition (c) of Definition 3.1 is violated. For example, $I_1 \cup (\mathbb{N} \setminus I_2) \notin \mathcal{F}$, while $I_1, \mathbb{N} \setminus I_2 \in \mathcal{F}$.

**3.19** Condition (a) of Definition 3.1 is satisfied because $\emptyset$, which is the complement of $\Omega$, is a finite set.

Now, take any $A \in \mathcal{F}$. Then, either $A$ is finite or its complement $\Omega \setminus A$ is finite. In the former case, $\Omega \setminus A \in \mathcal{F}$ because its complement $\Omega \setminus (\Omega \setminus A) = A$ is finite. In the latter case, $\Omega \setminus A \in \mathcal{F}$ because it is finite itself. This proves (b).

Let $A, B \in \mathcal{F}$. If both are finite sets, then $A \cup B$ is also finite, so $A \cup B \in \mathcal{F}$. If at least one of them, say, $A$ has a finite complement, then $A \cup B$ has a finite complement too because $\Omega \setminus (A \cup B) \subset \Omega \setminus A$ and $\Omega \setminus A$ is finite. Hence, $A \cup B \in \mathcal{F}$, completing the proof of (c).

**3.20** Both $\emptyset$ and $\mathbb{Z}$ are periodic sets. To show this for $\emptyset$, take $I = \emptyset$ and any $i$ in (3.1). For $\mathbb{Z}$, take $I = \{1, \dots, i\}$ and any $i$. This proves condition (a) of Definition 3.1.

If $A$ is a periodic set, then (3.1) holds for some $i$ and $I \subset \{1, \ldots, i\}$. Taking the same $i$ and replacing $I$ by $\{1, \ldots, i\} \setminus I$ in (3.1), we see immediately that $\mathbb{Z} \setminus A$ is also a periodic set, which proves (b).

Finally, let $A$ and $B$ be periodic sets; that is, $A$ is given by (3.1) for some positive integer $i$ and some $I \subset \{1, \ldots, i\}$, and $B$ is given by (3.1) with $i$ and $I$ replaced by some positive integer $j$ and some $J \subset \{1, \ldots, j\}$. Take $k = ij$ and

$$K = \left( \bigcup_{l=0}^{j-1} (I + li) \right) \cup \left( \bigcup_{l=0}^{i-1} (J + lj) \right).$$

Then, $K \subset \{1, \ldots, k\}$ and

$$A \cup B = \bigcup_{l=-\infty}^{+\infty} (K + lk),$$

which means that $A \cup B$ is a periodic set. Condition (c) has been verified.

**3.21** To verify that $\mathcal{F}_1$ and $\mathcal{F}_2$ are fields in the first place, see Problems 3.4 and 3.5.

The atoms in $\mathcal{F}_1$ are $\{1\}$, $\{2\}$, and $\{3, 4\}$. Each non-empty set in $\mathcal{F}_1$ contains at least one of these atoms.

The atoms in $\mathcal{F}_2$ are $(0, \frac{2}{3})$ and $[\frac{2}{3}, 1)$.

**3.22** The atoms are one-element subsets of $\Omega$.

Indeed, every one-element set $\{\omega\} \subset \Omega$ is an atom because it contains no other non-empty set, except itself.

Now, suppose that $A \subset \Omega$ has more than one element. Take any $\omega \in A$. Then, $\{\omega\} \subset A$, but $\{\omega\} \neq A$, so $A$ cannot be an atom. It follows that every atom in the field of all subsets of $\Omega$ must be a one-element set.

**3.23** Let $A$ and $B$ be two atoms. Then, $A \cap B \in \mathcal{F}$ by Problem 3.11. Since $A \cap B \subset A$ and $A$ is an atom, $A \cap B = \emptyset$ or $A \cap B = A$. Similarly, since $A \cap B \subset B$ and $B$ is an atom, $A \cap B = \emptyset$ or $A \cap B = B$. Thus, $A = A \cap B = B$ whenever $A \cap B \neq \emptyset$. It follows that if $A \neq B$, then $A \cap B = \emptyset$.

**3.24** Take any $\omega \in A$, which can be done because $A$ is non-empty. Since $\mathcal{F}$ is finite, there are finitely many events in $\mathcal{F}$ that contain $\omega$. Let us denote these events by $B_1, \ldots, B_n$. We claim that

$$B = B_1 \cap \cdots \cap B_n$$

is an atom contained in $A$.

First of all, $B \in \mathcal{F}$ by Problem 3.12. Next, $B \subset A$ because $\omega \in A$, so $A$ is among the sets $B_1, \dots, B_n$. Finally, suppose that $C$ is an event in $\mathcal{F}$ such that $C \subset B$. Then, either (i) $\omega \in C$ or (ii) $\omega \notin C$.

(i) If $\omega \in C$, then $C$ is among the sets $B_1, \dots, B_n$, so $B \subset C$. Since $C \subset B$, it follows that $B = C$.

(ii) If $\omega \notin C$, then $\omega \in B \setminus C$, which belongs to $\mathcal{F}$ by Problem 3.13. This means that $B \setminus C$ is among the sets $B_1, \dots, B_n$, so $B \subset B \setminus C$. But since $C \subset B$, the latter is possible only when $C = \emptyset$.

We have demonstrated that the only events contained in $B$ are the empty set $\emptyset$ and $B$ itself, which means that $B$ is an atom. This completes the proof.

**3.25** Because $\mathcal{F}$ is finite, there are finitely many atoms $A_1, \dots, A_k$ contained in $A$. We claim that $A = A_1 \cup \dots \cup A_k$.

Clearly, $A_1 \cup \dots \cup A_k \subset A$, since $A_1, \dots, A_k$ are contained in $A$. Suppose that $A \setminus (A_1 \cup \dots \cup A_k) \neq \emptyset$. Then, by 3.24, $A \setminus (A_1 \cup \dots \cup A_k)$ contains an atom $\tilde{A}$. It follows that $\tilde{A}$ is contained in $A$, but it is not among the atoms $A_1, \dots, A_k$, which is a contradiction. This proves that $A = A_1 \cup \dots \cup A_k$.

To show uniqueness, we take two sets of atoms $\{A_1, \dots, A_k\} \neq \{B_1, \dots, B_l\}$. Then, there is an atom $C$ belonging to one of these sets, but not to the other. Suppose that $C$ belongs to $\{A_1, \dots, A_k\}$, but not to $\{B_1, \dots, B_l\}$. Then, $C \subset A_1 \cup \dots \cup A_k$ and, by Problem 3.23, $C \cap B_i = \emptyset$ for each $i = 1, \dots, l$, which implies that $C \cap (B_1 \cup \dots \cup B_l) = \emptyset$. Because $C$ is non-empty, it follows that $A_1 \cup \dots \cup A_k \neq B_1 \cup \dots \cup B_l$, proving uniqueness.

**3.26** Since $\mathcal{F}$ is finite, so is the set $\mathcal{A}$ of all atoms in $\mathcal{F}$. We claim that if $n$ is the number of atoms, then $\mathcal{F}$ has $2^n$ elements, the same as the number of subsets of $\mathcal{A}$.

Indeed, to any set $\{A_1, \dots, A_k\}$ of atoms we can assign the event $A_1 \cup \dots \cup A_k \in \mathcal{F}$. By Problem 3.25, this defines a one-to-one correspondence between the subsets of $\mathcal{A}$ and the elements of $\mathcal{F}$, which proves the claim.

**3.27** There are five elements in $\mathcal{F}_1$, so it cannot be a field.

$\mathcal{F}_2$ has $8 = 2^3$ elements. The atoms are $[0, \frac{1}{4})$, $\{\frac{1}{4}\}$, and $(\frac{1}{4}, 1]$ and every set in $\mathcal{F}_2$ can be written as the union of some of these atoms. It follows that $\mathcal{F}_2$ is a field.

$\mathcal{F}_3$ has $8 = 2^3$ elements and there are three atoms $\{1\}$, $\{2\}$, and $\{3\}$. But the set $\{2, 3, 4\}$ cannot be written as the union of some of these atoms, so $\mathcal{F}_3$ is not a field.

**3.28** For every periodic set $A \neq \emptyset$, we shall construct a periodic set $B \subset A$ such that $\emptyset \neq B \neq A$.

Let $A$ be given by (3.1) for some positive integer $i$ and some $I \subset \{1, \dots, i\}$. Since $A \neq \emptyset$, it follows that $I \neq \emptyset$. We put

$$B = \bigcup_{l=-\infty}^{+\infty} (I + 2li),$$

which is clearly a periodic set. Then, $B \neq \emptyset$, since $I \subset B$. Because $I + li \subset A$ for every $l \in \mathbb{Z}$, we have $B \subset A$. But $(I + i) \cap B = \emptyset$, which means that $B \neq A$.

The above proves that there are no atoms in the field of periodic sets.

# 4

# Finitely Additive Probability

## 4.1 Theory and Problems

It was indicated in Chapter 1 that we need a function which would assign numbers to events, thus providing us with answers to questions such as "How likely is it that a given event $A$ takes place?" The basic idea encapsulated in condition (b) of Definition 4.1 is that to measure a set, we can decompose it into finitely many disjoint pieces, measure each piece separately, and then add up the results. In Chapter 6, this will be extended by allowing countably many pieces.

Suppose that $\Omega$ is a non-empty set.

▶ **Definition 4.1** Let $\mathcal{F}$ be a field of subsets of $\Omega$. We call a function $P : \mathcal{F} \to [0,1]$ a *finitely additive probability measure* if

(a) $P(\Omega) = 1$;

(b) (*additivity*) for any $A, B \in \mathcal{F}$ such that $A \cap B = \varnothing$

$$P(A \cup B) = P(A) + P(B).$$

We say that $P(A)$ is the *probability* of an event $A$.

Let us begin with some examples. Is $P$ a finitely additive probability measure?

**4.1** $\Omega = \{1, 2\}$, $\mathcal{F} = \{\varnothing, \Omega, \{1\}, \{2\}\}$,
$P(\varnothing) = 0$, $P(\Omega) = 1$, $P(\{1\}) = \frac{1}{3}$, $P(\{2\}) = \frac{2}{3}$.

**4.2** $\Omega = \{1,2,3,4\}$, $\mathcal{F} = \{\emptyset, \Omega, \{1,2\}, \{3,4\}\}$,
$P(\emptyset) = 0$, $P(\Omega) = 1$, $P(\{1,2\}) = \frac{5}{8}$, $P(\{3,4\}) = \frac{5}{8}$.

∘ **4.3** $\Omega = \{1,2,3,4\}$,
$\mathcal{F} = \{\emptyset, \Omega, \{1\}, \{4\}, \{2,3\}, \{1,4\}, \{1,2,3\}, \{2,3,4\}\}$,
$P(\emptyset) = 0$, $P(\Omega) = 1$, $P(\{1\}) = \frac{1}{8}$, $P(\{4\}) = \frac{3}{8}$, $P(\{2,3\}) = \frac{1}{2}$,
$P(\{1,4\}) = \frac{1}{2}$, $P(\{1,2,3\}) = \frac{5}{8}$, $P(\{2,3,4\}) = \frac{7}{8}$.

∘ **4.4** $\Omega = \{1,2,3,4\}$,
$\mathcal{F} = \{\emptyset, \Omega, \{1,2\}, \{3,4\}, \{3\}, \{4\}, \{1,2,3\}, \{1,2,4\}\}$,
$P(\emptyset) = 0$, $P(\Omega) = 1$, $P(\{1,2\}) = \frac{1}{2}$, $P(\{3,4\}) = \frac{1}{2}$,
$P(\{3\}) = \frac{1}{3}$, $P(\{4\}) = \frac{1}{6}$, $P(\{1,2,3\}) = \frac{5}{8}$, $P(\{1,2,4\}) = \frac{2}{3}$.

**4.5** Find $\alpha$ such that $P$ is a finitely additive probability measure, where $\Omega = \{1,2,3\}$, $\mathcal{F}$ consists of all subsets of $\Omega$, and $P(\{1\}) = \frac{1}{3}$, $P(\{2\}) = \frac{1}{6}$, $P(\{3\}) = \alpha$. Compute $P(\{1,2\})$, $P(\{1,3\})$, and $P(\{2,3\})$.

∘ **4.6** Find $\alpha$ such that $P$ is a finitely additive probability measure, where
$\Omega = \{1,2,3,4\}$,
$\mathcal{F} = \{\emptyset, \Omega, \{1\}, \{2,3\}, \{4\}, \{1,2,3\}, \{2,3,4\}, \{1,4\}\}$,
$P(\{1\}) = \frac{1}{4}$, $P(\{2,3\}) = \frac{1}{2}$, $P(\{4\}) = \alpha$.
Compute $P(\{1,2,3\})$, $P(\{2,3,4\})$, and $P(\{1,4\})$.

**4.7** Let $\Omega = \{1, \ldots, n\}$ and let $\mathcal{F}$ be the family of all subsets of $\Omega$ with $P(\{i\}) = \alpha_i$, $i = 1, \ldots, n$. Extend the function $P$ to a finitely additive probability measure defined on the whole field $\mathcal{F}$. What conditions have to be imposed on the numbers $\alpha_i$?

∘ **4.8** Let $\mathcal{F}$ be a finite field with atoms $A_1, \ldots, A_n$ and let $P(A_i) = \alpha_i$, $i = 1, \ldots, n$. Extend the function $P$ to a finitely additive probability measure defined on the whole field $\mathcal{F}$. What conditions have to be imposed on the numbers $\alpha_i$?

Prove the following facts (here $A$, $B$, $C$ are arbitrary events):

▶ **4.9** $P(\emptyset) = 0$.

▶ **4.10** If $A \subset B$, then $P(B \setminus A) = P(B) - P(A)$.

▶ **4.11** $P(\Omega \setminus A) = 1 - P(A)$.

▶ **4.12** If $A \subset B$, then $P(A) \leqslant P(B)$.

**4.13** If $A \cap B = \emptyset$, $A \cap C = \emptyset$, and $B \cap C = \emptyset$, then $P(A \cup B \cup C) = P(A) + P(B) + P(C)$.

▶ **4.14** If $A_1, \ldots, A_n$ are pairwise disjoint (i.e., $A_i \cap A_j = \emptyset$ for $i \neq j$), then

$$P\left(\bigcup_{i=1}^{n} A_i\right) = \sum_{i=1}^{n} P(A_i).$$

In the problems to follow next, we do not assume that the sets in question are pairwise disjoint. The property in Problems 4.15 and 4.17 is known as *subadditivity*.

▶ **4.15** Show that

$$P(A \cup B) \leqslant P(A) + P(B).$$

**4.16** Give an example such that $P(A \cup B) < P(A) + P(B)$.

**4.17** Show that

$$P\left(\bigcup_{i=1}^{n} A_i\right) \leqslant \sum_{i=1}^{n} P(A_i).$$

▶ **4.18** Show that

$$P(A \cup B) = P(A) + P(B) - P(A \cap B).$$

**4.19** Find an expression for $P(A \cup B \cup C)$ along the lines of Problem 4.18.

∗ **4.20** Generalize Problem 4.18 by finding an expression for $P(A_1 \cup \cdots \cup A_n)$.

**4.21** Suppose that $P(A) = \frac{1}{2}$ and $P(B) = \frac{2}{3}$. Show that $\frac{1}{6} \leqslant P(A \cap B) \leqslant \frac{1}{2}$. Give examples to demonstrate that the extreme values $\frac{1}{6}$ and $\frac{1}{2}$ can be attained.

**4.22** Suppose that $P(A) = \frac{1}{2}$ and $P(B) = \frac{1}{3}$. Show that $\frac{1}{2} \leqslant P(A \cup B) \leqslant \frac{5}{6}$. Give examples to show that the extreme values $\frac{1}{2}$ and $\frac{5}{6}$ can be attained.

**4.23** Show that $P(A \triangle B) = P(A) + P(B) - 2P(A \cap B)$.

**4.24** Show that $|P(A) - P(B)| \leqslant P(A \triangle B)$.

**4.25** Verify the inequality $P(A \triangle C) \leqslant P(A \triangle B) + P(B \triangle C)$.

**4.26** Prove that if $P(A) = P(B) = 0$, then $P(A \cup B) = 0$.

**4.27** Prove that if $P(A) = P(B) = 1$, then $P(A \cap B) = 1$.

Finally, let us consider examples of finitely additive probability measures defined on an infinite field of sets.

**4.28** Let $\Omega$ be an infinite set and $\mathcal{F}$ the field of finite subsets of $\Omega$ and their complements introduced in Problem 3.19. Show that

$$P(A) = \begin{cases} 0 & \text{if } A \text{ is a finite set} \\ 1 & \text{if } A \text{ has a finite complement} \end{cases}$$

is a finitely additive probability measure.

$*$ **4.29** For any periodic set $A$ as defined in Problem 3.20, we put

$$P(A) = \frac{\#I}{i} \tag{4.1}$$

if $A$ can be represented by (3.1) for some positive integer $i$ and some $I \subset \{1, \ldots, i\}$. Show that $P$ is a finitely additive probability measure on the field of periodic sets.

## 4.2 Hints

**4.1** The choice of pairwise disjoint events $A$ and $B$ is very limited. The empty set does not contribute to either side of the equality in condition (b) of Definition 4.1. The full set $\Omega$ has a non-empty intersection with any set except the empty one. We only need to consider the remaining events $\{1\}$ and $\{2\}$.

**4.2** Condition (b) of Definition 4.1 does not hold. Find the appropriate sets $A$ and $B$.

**4.3** See Problem 4.1.

**4.4** See Problem 4.1.

**4.5** Since $P(\Omega) = 1$, the given measures should add up to 1, providing an equation for $\alpha$.

**4.6** See Problem 4.5.

**4.7** The numbers $\alpha_i$ are some of the values of the function $P$. First, consider the question: What are the admissible values of $P$? This gives certain restrictions on $\alpha_i$. Next, take into account that $P(\Omega) = 1$.

To extend $P$ to all subsets of $\Omega$, give a formula for $P(A)$, where $A \subset \Omega$ has two elements, say $A = \{1, 2\}$, and then generalize it. Use condition (b) of Definition 4.1. Try to make the notation as simple as possible.

**4.8** In Problem 4.7, use $A_i$ in place of $\{i\}$.

**4.9** Use condition (b) of Definition 4.1 with $A = \emptyset$ and $B = \Omega$.

**4.10** Consider the two disjoint sets $A$ and $B \setminus A$ and apply condition (b) of Definition 4.1.

**4.11** See Problem 4.10.

**4.12** What do you need to know about the difference $P(B) - P(A)$ to obtain the desired inequality? Can you compute the difference using one of the previous exercises?

**4.13** Use condition (b) of Definition 4.1 twice.

**4.14** Use condition (b) of Definition 4.1 repeatedly or apply mathematical induction.

**4.15** Represent $A \cup B$ as the sum of two disjoint sets and apply Problem 4.12. If one of these disjoint sets is $B$, what is the other set?

**4.16** $P(A \cup B)$ cannot be greater than 1, but it is easy to find $A$ and $B$ such that $P(A) + P(B) > 1$.

**4.17** Use Problem 4.15 and induction on $n$.

**4.18** Figure 4.1 is a diagram of two overlapping events.

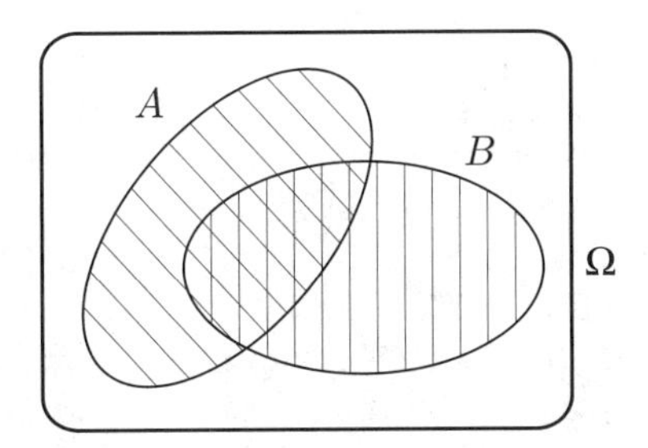


FIGURE 4.1. Two overlapping events $A$ and $B$

The union $A \cup B$ can be decomposed into three disjoint bits. Use condition (b) of Definition 4.1 (in fact, Problem 4.13 or 4.14 for $n = 3$). Do the same with $A$ and $B$ (each splits into two pieces) and substitute the results to the equality in question.

The intuition related to measuring areas is useful. The number $P(A) + P(B)$ is too large for the measure of $A \cup B$, since the intersection has been covered twice, so we have to subtract the measure of the intersection to compensate.

**4.19** A diagram of three overlapping events (Figure 4.2) may be of some help.

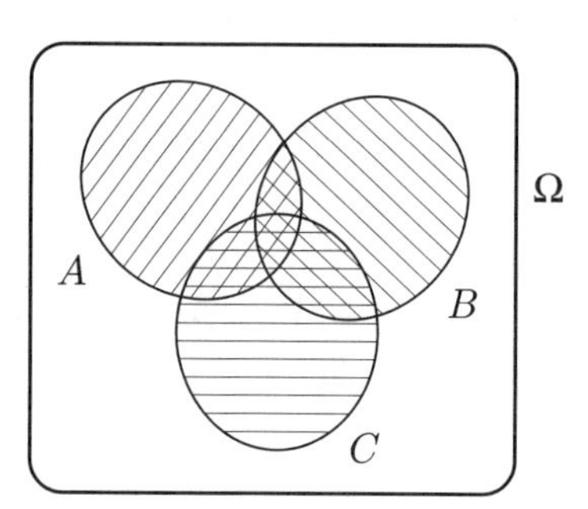


FIGURE 4.2. Three overlapping events $A$, $B$, and $C$

Treating $A \cup B$ as one set and $C$ as the other one, you can apply Problem 4.18. Then, use it again to compute $P(A \cup B)$. You will have to use it once more to deal with one of the components.

The intuitive idea that measuring the probability is analogous to measuring the area can be helpful: The sum of all probabilities is too large (some bits are measured twice and one bit even three times) and we have to compensate for it.

**4.20** Try to guess the general formula. Think of bits that are measured once, twice, three times, ..., $n$ times in the sum $P(A_1)+\cdots+P(A_n)$. Once you arrive at the desired formula, prove it by induction on $n$. It helps to design some concise notation because the general formula tends to be quite long.

**4.21** To obtain the upper estimate, use the fact that $A \cap B \subset B$. What does it tell you about probabilities? To obtain the lower estimate, express the probability of $A \cup B$ by those of $A$, $B$, and $A \cap B$ and take into account that $P(A \cup B) \leqslant 1$.

To show that the extreme values can be attained, look for examples where $B \subset A$ for the upper bound and $A \cup B = \Omega$ for the lower bound.

**4.22** You can follow the solution to Problem 4.21 to obtain estimates for the probability of $(\Omega \setminus A) \cap (\Omega \setminus B)$ and use the fact that $A \cap B$ is its complement. But there is also a simple direct argument.

To show that the lower and upper bounds can be attained, look for examples in which $A \subset B$ and $A \cap B = \emptyset$, respectively.

**4.23** First, show that $P(A \setminus B) = P(A) - P(A \cap B)$.

**4.24** Can $P(A)-P(B)$ be greater than $P(A \setminus B)$? Can $P(A \setminus B)$ be greater than $P(A \triangle B)$?

**4.25** Sketch a diagram showing three overlapping sets $A$, $B$, and $C$ (or refer to Figure 4.2) to check that $A \triangle C$ is contained in $(A \triangle B) \cup (B \triangle C)$. What does it imply for probabilities?

**4.26** Use subadditivity.

**4.27** Use Problem 4.26 for the complements of $A$ and $B$.

**4.28** When verifying condition (b) of Definition 4.1, observe that if $A, B \in \mathcal{F}$ are disjoint events, then one or both of them must be finite.

**4.29** First of all, show that $P(A)$ is well defined, i.e., independent of the choice of $i$ and $I$ in (3.1).

To verify condition (a) of Definition 4.1, you can choose any period $i$ when writing $A = \mathbb{Z}$ in the form (3.1). What is the simplest choice of $i$? What does the corresponding set $I$ look like? How many elements does it have?

To verify (b), choose a common period $i$ for $A$ and $B$ in the representation (3.1). (For the meaning of "period," see Hint 3.20.)

## 4.3 Solutions

**4.1** We check condition (b) of Definition 4.1. The empty set does not contribute to either side of the equality in (b):

$$P(A \cup \emptyset) = P(A) = P(A) + 0 = P(A) + P(\emptyset).$$

The full set $\Omega$ is not disjoint with any set except the empty one, and this is covered by the above argument. We only need to consider the case when $A = \{1\}$ and $B = \{2\}$. Then, $P(A \cup B) = P(\Omega) = 1$, which is equal to $P(A)+P(B)$. So $P$ is a finitely additive probability measure.

**4.2** Let $A = \{1,2\}$ and $B = \{3,4\}$. Clearly, $A \cap B = \emptyset$ and $P(A \cup B) = P(\Omega) = 1$, but $P(A) + P(B) = \frac{10}{8}$. So, condition (b) does not hold and $P$ is not a finitely additive probability measure.

**4.5** Since $1 = P(\Omega) = P(\{1\}) + P(\{2\}) + P(\{3\}) = \frac{1}{3} + \frac{1}{6} + \alpha$, we find that $\alpha = \frac{1}{2}$. Then,

$$P(\{1,2\}) = P(\{1\}) + P(\{2\}) = \frac{1}{2},$$

$$P(\{1,3\}) = P(\{1\}) + P(\{3\}) = \frac{5}{6},$$

$$P(\{2,3\}) = P(\{2\}) + P(\{3\}) = \frac{2}{3}.$$

**4.7** First, since the values of $P$ must be in $[0, 1]$, we have to assume that

$$0 \leqslant \alpha_i \leqslant 1.$$

Next,

$$1 = P(\Omega) = \sum_{i=1}^{n} P(\{i\}),$$

so

$$\sum_{i=1}^{n} \alpha_i = 1.$$

For any $A \subset \Omega$, we put

$$P(A) = \sum_{i \in A} P(\{i\}) = \sum_{i \in A} \alpha_i.$$

**4.9** Let $A = \emptyset$ and $B = \Omega$. Clearly, they are disjoint: $\emptyset \cap \Omega = \emptyset$. Condition (b) of Definition 4.1 gives

$$P(\Omega) = P(\emptyset \cup \Omega) = P(\emptyset) + P(\Omega).$$

It follows that $1 = P(\emptyset) + 1$, so $P(\emptyset) = 0$.

**4.10** Note that $B = A \cup (B \setminus A)$, where $A$ and $B \setminus A$ are disjoint, so $P(A) + P(B \setminus A) = P(A \cup (B \setminus A)) = P(B)$. Subtracting $P(A)$ from both sides, we obtain $P(B \setminus A) = P(B) - P(A)$, as required.

**4.11** Since $\Omega = A \cup (\Omega \setminus A)$, $1 = P(\Omega) = P(A) + P(\Omega \setminus A)$.
Here, we have a particular case of Problem 4.10 with $B = \Omega$.

**4.12** $P(B) - P(A) = P(B \setminus A) \geqslant 0$, using Problem 4.10.

**4.13** First, by (b) of Definition 4.1,

$$P(A \cup B) = P(A) + P(B).$$

Next, by (b) again, for the two sets $D = A \cup B$ and $C$, we have

$$P(A \cup B \cup C) = P(D \cup C) = P(D) + P(C) = P(A \cup B) + P(C).$$

Note that we may apply (b) to $D$ and $C$ because

$$D \cap C = (A \cup B) \cap C = (A \cap C) \cup (B \cap C) = \emptyset.$$

Now, substitute to get the result:

$$P(A \cup B \cup C) = P(A \cup B) + P(C) = P(A) + P(B) + P(C).$$

**4.14** We argue by induction. For $n = 1$, there is nothing to prove. Induction hypothesis: Suppose that the claim is valid for some $n = k$,

$$P\left(\bigcup_{i=1}^{k} A_i\right) = \sum_{i=1}^{k} P(A_i).$$

Consider $k+1$ and use condition (b) of Definition 4.1 and the induction hypothesis:

$$\begin{aligned}
P\left(\bigcup_{i=1}^{k+1} A_i\right) &= P\left(\left(\bigcup_{i=1}^{k} A_i\right) \cup A_{k+1}\right) \\
&= P\left(\bigcup_{i=1}^{k} A_i\right) + P(A_{k+1}) && \text{by condition (b)} \\
&= \sum_{i=1}^{k} P(A_i) + P(A_{k+1}) && \text{by the induction hypothesis} \\
&= \sum_{i=1}^{k+1} P(A_i).
\end{aligned}$$

**4.15** Since $A \cup B = (A \setminus B) \cup B$ and $(A \setminus B) \cap B = \emptyset$, it follows that

$$P(A \cup B) = P(A \setminus B) + P(B) \leqslant P(A) + P(B).$$

The inequality holds because $A \setminus B \subset A$ and Problem 4.12 implies that $P(A \setminus B) \leqslant P(A)$.

**4.16** Take $A = B = \Omega$. Then, $P(A \cup B) = P(\Omega) = 1$ and $P(A) + P(B) = 2P(\Omega) = 2$; that is, $P(A \cup B) < P(A) + P(B)$.

**4.17** We prove the inequality by induction on $n$. For $n = 1$, it is trivial. Induction hypothesis: Suppose that for some $n = k$,

$$P\left(\bigcup_{i=1}^{k} A_i\right) \leqslant \sum_{i=1}^{k} P(A_i).$$

Then, taking $k+1$ events $A_1, \ldots, A_k, A_{k+1}$, we have

$$\begin{aligned} P\left(\bigcup_{i=1}^{k+1} A_i\right) &= P\left(\left(\bigcup_{i=1}^{k} A_i\right) \cup A_{k+1}\right) \\ &\leqslant P\left(\bigcup_{i=1}^{k} A_i\right) + P(A_{k+1}) && \text{by Problem 4.15} \\ &\leqslant \sum_{i=1}^{k} P(A_i) + P(A_{k+1}) && \text{by the induction hypothesis} \\ &= \sum_{i=1}^{k+1} P(A_i). \end{aligned}$$

The equality has been verified for $n = k+1$, completing the proof.

**4.18** Compute the left-hand side: We decompose $A \cup B$ into three pairwise disjoint pieces

$$A \cup B = (A \setminus B) \cup (B \setminus A) \cup (A \cap B)$$

and use condition (b) of Definition 4.1 to get

$$P(A \cup B) = P(A \setminus B) + P(B \setminus A) + P(A \cap B).$$

Now, compute the right-hand side. Since $A \cap B \subset A$ and $A \cap B \subset B$, by Problem 4.10 it follows that

$$P(A \setminus B) = P(A) - P(A \cap B), \quad P(B \setminus A) = P(B) - P(A \cap B).$$

Substituting these for $P(A \setminus B)$ and $P(B \setminus A)$, we get the result.

An alternative, more elegant proof goes as follows: Since $A \cup B = (A \setminus (A \cap B)) \cup B$ and $(A \setminus (A \cap B)) \cap B = \emptyset$,

$$\begin{aligned} P(A \cup B) &= P(A \setminus (A \cap B)) + P(B) && \text{by condition (b)} \\ &= P(A) - P(A \cap B) + P(B) && \text{by Problem 4.10.} \end{aligned}$$

**4.19** We use Problem 4.18, obtaining

$$\begin{aligned} P(A \cup B \cup C) &= P((A \cup B) \cup C) \\ &= P(A \cup B) + P(C) - P((A \cup B) \cap C). \end{aligned}$$

Now,

$$P(A \cup B) = P(A) + P(B) - P(A \cap B)$$

and

$$\begin{aligned} P((A \cup B) \cap C) &= P((A \cap C) \cup (B \cap C)) \\ &= P(A \cap C) + P(B \cap C) - P((A \cap C) \cap (B \cap C)) \\ &= P(A \cap C) + P(B \cap C) - P(A \cap B \cap C), \end{aligned}$$

again by Problem 4.18. Finally,

$$\begin{aligned} P(A \cup B \cup C) = \; & P(A) + P(B) + P(C) \\ & -P(A \cap B) - P((A \cap C) - (B \cap C)) \\ & +P(A \cap B \cap C). \end{aligned}$$

**4.20** The desired formula reads

$$\begin{aligned} P(A_1 \cup \cdots \cup A_n) = & \sum_i P(A_i) - \sum_{i<j} P(A_i \cap A_j) \\ & + \sum_{i<j<k} P(A_i \cap A_j \cap A_k) - \cdots \\ & +(-1)^{n+1} P(A_1 \cap \cdots \cap A_n), \end{aligned}$$

where all the summation indices $i, j, k, \ldots$ belong to $\{1, \ldots, n\}$. This is known as the *inclusion-exclusion formula.* It can also be written as

$$P\left(\bigcup_{i \in I_n} A_i\right) = \sum_{\emptyset \neq I \subset I_n} (-1)^{\#I+1} P\left(\bigcap_{i \in I} A_i\right),$$

where $I_n = \{1, \ldots, n\}$. The sum on the right-hand side is over all non-empty subsets $I$ of $I_n$ and, as usual, $\#I$ denotes the number of elements of $I$.

We shall prove the formula by induction on $n$. For $n = 1$, it is obvious. Induction hypothesis: Suppose that for some $n = k$,

$$P\left(\bigcup_{i \in I_k} A_i\right) = \sum_{\emptyset \neq I \subset I_k} (-1)^{\#I+1} P\left(\bigcap_{i \in I} A_i\right).$$

Then,

$$
\begin{aligned}
P\left(\bigcup_{i\in I_{k+1}} A_i\right) &= P\left(\left(\bigcup_{i\in I_k} A_i\right)\cup A_{k+1}\right)\\
&= P\left(\bigcup_{i\in I_k} A_i\right) + P(A_{k+1}) - P\left(\left(\bigcup_{i\in I_k} A_i\right)\cap A_{k+1}\right)\\
&\qquad\qquad\text{by Problem 4.18}\\
&= \sum_{\emptyset\neq I\subset I_k} (-1)^{\#I+1} P\left(\bigcap_{i\in I} A_i\right) + P(A_{k+1}) \quad\text{using the hypothesis}\\
&\quad -P\left(\bigcup_{i\in I_k} (A_i\cap A_{k+1})\right) \quad\text{and rearranging the last term}\\
&= \sum_{\emptyset\neq I\subset I_k} (-1)^{\#I+1} P\left(\bigcap_{i\in I} A_i\right) + P(A_{k+1})\\
&\quad - \sum_{\emptyset\neq I\subset I_k} (-1)^{\#I+1} P\left(\bigcap_{i\in I} A_i\cap A_{k+1}\right) \quad\text{also by the hypothesis}\\
&= \sum_{\emptyset\neq I\subset I_{k+1}} (-1)^{\#I+1} P\left(\bigcap_{i\in I} A_i\right) \quad\text{collecting all terms in one sum.}
\end{aligned}
$$

This means that the formula holds for $n = k+1$. The proof is completed. See Solution 9.17 for another proof of this formula.

**4.21** Since $A\cap B\subset A$,

$$P(A\cap B)\leqslant P(A)=\frac{1}{2}.$$

Next, from Problem 4.18,

$$1\geqslant P(A\cup B)=P(A)+P(B)-P(A\cap B),$$

so

$$P(A\cap B)\geqslant P(A)+P(B)-1=\frac{1}{2}+\frac{2}{3}-1=\frac{1}{6}.$$

To show that the extreme values can actually be attained, we take $\Omega=\{1,2,3,4,5,6\}$ with uniform probability measure $P$. Then, putting, for example,

$$A=\{1,2,3\},\quad B=\{1,2,3,4\},$$

we have

$$P(A) = \frac{1}{2}, \quad P(B) = \frac{2}{3}, \quad P(A \cap B) = \frac{1}{2}.$$

If we take

$$A = \{1, 2, 3\}, \quad B = \{3, 4, 5, 6\},$$

then

$$P(A) = \frac{1}{2}, \quad P(B) = \frac{2}{3}, \quad P(A \cap B) = \frac{1}{6}.$$

**4.22** Since $A \subset A \cup B$, we obtain the lower estimate:

$$\frac{1}{2} = P(A) \leqslant P(A \cup B).$$

The upper estimate holds because

$$\begin{aligned} P(A \cup B) &= P(A) + P(B) - P(A \cap B) \\ &\leqslant P(A) + P(B) = \frac{1}{2} + \frac{1}{3} = \frac{5}{6}. \end{aligned}$$

To show that the extreme values can be attained, we take $\Omega = \{1, 2, 3, 4, 5, 6\}$ with uniform probability measure $P$. For

$$A = \{1, 2, 3\}, \quad B = \{1, 2\},$$

we have

$$P(A) = \frac{1}{2}, \quad P(B) = \frac{1}{3}, \quad P(A \cup B) = \frac{1}{2}.$$

If we take

$$A = \{1, 2, 3\}, \quad B = \{4, 5\},$$

then

$$P(A) = \frac{1}{2}, \quad P(B) = \frac{1}{3}, \quad P(A \cup B) = \frac{5}{6}.$$

**4.23** Because $A \triangle B = (A \setminus B) \cup (B \setminus A)$ and $(A \setminus B) \cap (B \setminus A) = \emptyset$, we have

$$\begin{aligned} P(A \triangle B) &= P(A \setminus B) + P(B \setminus A) \\ &= P(A \setminus (A \cap B)) + P(B \setminus (A \cap B)) \\ &= P(A) - P(A \cap B) + P(B) - P(A \cap B) \\ &= P(A) + P(B) - 2P(A \cap B). \end{aligned}$$

**4.24** $P(A) - P(B)$ can be estimated as follows:

$$\begin{aligned} P(A) - P(B) &\leqslant P(A \cup B) - P(B) && \text{because } A \subset A \cup B \\ &= P(A \setminus B) && \text{because } B \subset A \cup B \\ &\leqslant P(A \setminus B) + P(B \setminus A) && \text{because } 0 \leqslant P(B \setminus A) \\ &= P(A \triangle B) && \text{by (b) of Definition 4.1.} \end{aligned}$$

Interchanging $A$ and $B$, we obtain $P(B) - P(A) \leqslant P(A \triangle B)$. It follows that $|P(A) - P(B)| \leqslant P(A \triangle B)$.

**4.25** Refer to Figure 4.2, which shows three overlapping events $A$, $B$, and $C$. You will see immediately that

$$A \triangle C \subset (A \triangle B) \cup (B \triangle C).$$

It follows that

$$\begin{aligned} P(A \triangle C) &\leqslant P((A \triangle B) \cup (B \triangle C)) \\ &\leqslant P(A \triangle B) + P(B \triangle C). \end{aligned}$$

**4.26** By Problem 4.17,

$$P(A \cup B) \leqslant P(A) + P(B) = 0,$$

so $P(A \cup B) = 0$.

**4.27** Since $P(A) = P(B) = 1$, it follows that $P(\Omega \setminus A) = P(\Omega \setminus B) = 0$. By Problem 4.26,

$$P((\Omega \setminus A) \cup (\Omega \setminus B)) = 0.$$

But $A \cap B$ is the complement of $(\Omega \setminus A) \cup (\Omega \setminus B)$, so $P(A \cap B) = 1$ by Problem 4.11.

**4.28** The complement of $\Omega$ is the empty set $\emptyset$, which is finite, so $P(\Omega) = 1$. Condition (a) of Definition 4.1 is satisfied.

To verify condition (b), take two events $A, B \in \mathcal{F}$ such that $A \cap B = \emptyset$. By de Morgan's law,

$$(\Omega \setminus A) \cup (\Omega \setminus B) = \Omega \setminus (A \cap B).$$

Since $\Omega$ is an infinite set and $A \cap B = \emptyset$, it follows that $\Omega \setminus A$ or $\Omega \setminus B$ must be infinite. But because $A, B \in \mathcal{F}$, this means that $A$ or $B$ must be finite. If $A$ is finite and $B$ has a finite complement, then $A \cup B$ has a finite complement. In this case, $P(A) = 0$ and $P(B) = P(A \cup B) = 1$, so (b) is satisfied. If both $A$ and $B$ are finite, then $A \cup B$ is also finite. In this case, $P(A) = P(B) = P(A \cup B) = 0$, which implies that (b) is also satisfied.

**4.29** First, we must verify that $P$ is well defined on periodic sets; that is, $P(A)$ given by (4.1) is independent of the choice of $i$ and $I$ as long as (3.1) is satisfied.

Suppose that $A \subset \mathbb{Z}$ can be represented in two ways:

$$A = \bigcup_{l=-\infty}^{\infty} (I + li) = \bigcup_{l=-\infty}^{\infty} (J + lj),$$

where $i$ and $j$ are positive integers and

$$I \subset \{1, \dots, i\}, \qquad J \subset \{1, \dots, j\}.$$

If we take

$$K = A \cap \{1, \dots, ij\},$$

then

$$K = \bigcup_{l=1}^{j} (I + li) = \bigcup_{l=1}^{i} (J + lj);$$

hence,

$$\#K = j(\#I) = i(\#J)$$

because the sets $I + i, I + 2i, \dots, I + ji$ are pairwise disjoint and so are the sets $J + j, J + 2j, \dots, J + ij$. Therefore,

$$\frac{\#I}{i} = \frac{j(\#I)}{ij} = \frac{\#K}{ij} = \frac{i(\#J)}{ij} = \frac{\#J}{j}.$$

This proves that $P$ is well defined on periodic sets.

To show that $P$ is a finitely additive probability measure, we shall verify conditions (a) and (b) of Definition 4.1. Taking $i = 1$ and $I = \{1\}$, we obtain a representation (3.1) of $\Omega = \mathbb{Z}$. Hence,

$$P(\Omega) = \frac{\#\{1\}}{1} = 1,$$

which proves condition (a).

Now, let $A$ and $B$ be two disjoint periodic sets such that

$$A = \bigcup_{l=-\infty}^{\infty} (I + li), \qquad B = \bigcup_{l=-\infty}^{\infty} (J + lj),$$

where $i$ and $j$ are positive integers and

$$I \subset \{1, \dots, i\}, \qquad J \subset \{1, \dots, j\}.$$

Putting

$$M = \bigcup_{l=1}^{j} (I + li), \qquad N = \bigcup_{l=1}^{i} (J + lj),$$

we can write $A$ and $B$ as

$$A = \bigcup_{l=-\infty}^{\infty} (M + lij), \qquad B = \bigcup_{l=-\infty}^{\infty} (N + lij)$$

and $A \cup B$ as

$$A \cup B = \bigcup_{l=-\infty}^{\infty} (M \cup N + lij).$$

Since $M = A \cap \{1, \ldots, ij\}$, $N = B \cap \{1, \ldots, ij\}$, and $A \cap B = \emptyset$, it follows that $M \cap N = \emptyset$. Therefore,

$$\begin{aligned} P(A \cup B) &= \frac{\#(M \cup N)}{ij} = \frac{\#M + \#N}{ij} \\ &= \frac{j(\#I) + i(\#J)}{ij} = \frac{\#I}{i} + \frac{\#J}{j} = P(A) + P(B), \end{aligned}$$

which proves condition (b).

# 5

# Sigma Fields

## 5.1 Theory and Problems

In this section, we consider families of sets satisfying conditions similar to but stronger than those imposed on fields. Because computing probability often involves the approximation of events by sequences of simpler ones, we need to ensure that the union of a sequence of events is always an event.

▶ **Definition 5.1** A family $\mathcal{F}$ of subsets of a non-empty set $\Omega$ is called a *$\sigma$-field* (sigma-field) on $\Omega$ if

(a) $\Omega \in \mathcal{F}$;

(b) if $A \in \mathcal{F}$, then $\Omega \setminus A \in \mathcal{F}$;

(c) if $A_1, A_2, \ldots \in \mathcal{F}$, then $\bigcup_{i=1}^{\infty} A_i \in \mathcal{F}$.

The elements of a $\sigma$-field are called *events*.

The conditions of this definition can be read as follows:

(a) the set of all outcomes is an event;

(b) the complement of an event is an event;

(c) the union of a sequence of events is an event.

**5.1** Show that a $\sigma$-field is also a field.

**5.2** If a field $\mathcal{F}$ is finite, then it is also a $\sigma$-field.

**5.3** Show that the family of all subsets of $\Omega$ is a $\sigma$-field.

**5.4** Is the family $\mathcal{F}$ consisting of all finite subsets of $\Omega$ and their complements always a $\sigma$-field?

**5.5** Is the family $\mathcal{F}$ consisting of all countable subsets of $\Omega$ and their complements always a $\sigma$-field?

**5.6** Let $\mathcal{F}$ be a $\sigma$-field on $\Omega = [0,1]$ such that $[\frac{1}{n+1}, \frac{1}{n}] \in \mathcal{F}$ for $n = 1, 2, \ldots$ . Show that

a) $\{0\} \in \mathcal{F}$,
b) $\{\frac{1}{n} : n = 2, 3, \ldots\} \in \mathcal{F}$,
c) $(\frac{1}{n}, 1] \in \mathcal{F}$ for all $n$,
d) $(0, \frac{1}{n}] \in \mathcal{F}$ for all $n$.

▶ **5.7** Let $\mathcal{F}$ be a $\sigma$-field. Demonstrate that if $A_1, A_2, \ldots \in \mathcal{F}$, then

$$\bigcap_{i=1}^{\infty} A_i \in \mathcal{F}.$$

▶ **5.8** Let $\Omega$ and $\tilde{\Omega}$ be arbitrary sets and let $X : \tilde{\Omega} \to \Omega$ be any function. Show that if $\mathcal{F}$ is a $\sigma$-field on $\Omega$, then $\tilde{\mathcal{F}} = \{X^{-1}(A) : A \in \mathcal{F}\}$ is a $\sigma$-field on $\tilde{\Omega}$.

**5.9** Let $\mathcal{F}$ be a $\sigma$-field on $\Omega$ and let $A \subset \Omega$. Show that $\tilde{\mathcal{F}} = \{A \cap B : B \in \mathcal{F}\}$ is a $\sigma$-field on $A$.

▶ **5.10** Let $\tilde{\Omega}$ and $\Omega$ be arbitrary sets and let $X : \tilde{\Omega} \to \Omega$ be any function. Show that if $\tilde{\mathcal{F}}$ is a $\sigma$-field on $\tilde{\Omega}$, then $\mathcal{F} = \{A \subset \Omega : X^{-1}(A) \in \tilde{\mathcal{F}}\}$ is a $\sigma$-field on $\Omega$.

∗ **5.11** Find an example of a function $X : \tilde{\Omega} \to \Omega$ and a $\sigma$-field $\tilde{\mathcal{F}}$ on $\tilde{\Omega}$ such that $\{X(A) : A \in \tilde{\mathcal{F}}\}$ is not a $\sigma$-field.

∗ **5.12** Is the field of periodic subsets of $\mathbb{Z}$ in Problem 3.20 a $\sigma$-field?

∗ **5.13** Find two $\sigma$-fields such that their union is not a $\sigma$-field.

**5.14** Prove that if $\mathcal{F}_1$ and $\mathcal{F}_2$ are $\sigma$-fields on $\Omega$, then $\mathcal{F}_1 \cap \mathcal{F}_2$ is also a $\sigma$-field.

▶ **5.15** Show that if $\{\mathcal{F}_j\}_{j \in J}$ is any collection of $\sigma$-fields defined on the same set $\Omega$, then their intersection $\bigcap_{j \in J} \mathcal{F}_j$ is also a $\sigma$-field.

In many random situations, the outcome is a real number. We may be interested in the probability that it belongs to a given interval $(a, b)$. In a mathematical model, we can identify such an event with the interval $(a, b)$ itself. To consider all such events, we need a $\sigma$-field of subsets of the real line $\mathbb{R}$ containing all intervals. However, the family of all intervals does not form a $\sigma$-field (for example, $(0, 1) \cup (2, 3)$ is not an interval), so we have to enlarge it in some way. The simplest solution would be to take the family of all subsets of $\mathbb{R}$, but this $\sigma$-field is too large for our purposes (see the remark following Definition 6.2). We need another way of extending the family of intervals to a $\sigma$-field.

In Problems 3.6 and 3.7, we came across the idea of extending a given family of sets to a field by gradually building up. However, in the case of a $\sigma$-field, we have no guarantee that this process will end after finitely many steps if the given family of sets is infinite. The following definition employs a different idea: the intersection of all $\sigma$-fields containing the given family of sets. This family of $\sigma$-fields is non-empty (see Problem 5.3) and its intersection is a $\sigma$-field (see Problem 5.15).

▶ **Definition 5.2** For a family $\mathcal{A}$ of subsets of $\Omega$, we put

$$\mathcal{F}_{\mathcal{A}} = \bigcap \{\mathcal{F} : \mathcal{F} \text{ is a } \sigma\text{-field such that } \mathcal{A} \subset \mathcal{F}\}$$

and call it the *$\sigma$-field generated by* $\mathcal{A}$.

**5.16** Let $\mathcal{A} = \{[0,1]\} \cup \{[\frac{1}{2^{n+1}}, \frac{1}{2^n}) : n = 0,1,2,\dots\}$. Which of the following sets belong to $\mathcal{F}_{\mathcal{A}}$:

a) $\{0\}$, d) $\{\frac{1}{3}\}$, g) $[0, \frac{1}{2}]$,
b) $\{1\}$, e) $\{0,1\}$, h) $[\frac{1}{4}, 1)$,
c) $\{\frac{1}{2}\}$, f) $(\frac{1}{4}, 1]$, i) $(0, \frac{1}{2})$?

**5.17** Show that $\mathcal{F}_{\mathcal{A}} = \mathcal{A}$ if and only if $\mathcal{A}$ is a $\sigma$-field.

The next problem explains why $\mathcal{F}_{\mathcal{A}}$ is often referred to as the *smallest* $\sigma$-field containing $\mathcal{A}$.

▶ **5.18** Let $\tilde{\mathcal{F}}$ be a $\sigma$-field that satisfies the conditions

(a) $\mathcal{A} \subset \tilde{\mathcal{F}}$;

(b) if $\mathcal{F}$ is a $\sigma$-field and $\mathcal{A} \subset \mathcal{F}$, then $\tilde{\mathcal{F}} \subset \mathcal{F}$.

Show that $\mathcal{F}_{\mathcal{A}} = \tilde{\mathcal{F}}$.

**5.19** Let $\mathcal{A}$ be the family of all finite subsets of $\Omega$. Show that $\mathcal{F}_{\mathcal{A}}$ is the $\sigma$-field consisting of all countable subsets of $\Omega$ and their complements introduced in Problem 5.5.

∗ **5.20** If $\mathcal{C} \subset \mathcal{D}$, then $\mathcal{F}_{\mathcal{C}} \subset \mathcal{F}_{\mathcal{D}}$.

▶ **Definition 5.3** When $\Omega = \mathbb{R}$ and $\mathcal{A}$ is the family of all open intervals, then we call $\mathcal{F}_{\mathcal{A}}$ the $\sigma$-field of *Borel sets* and denote it by $\mathcal{B}$.

**5.21** Show that the following are Borel sets (that is, belong to $\mathcal{B}$):

a) $(a,b)$, f) any finite set,
b) $(a,+\infty)$, g) any countable set,
c) $(-\infty, a)$, h) the set $\mathbb{N}$ of natural numbers,
d) $[a,b]$, i) the set $\mathbb{Q}$ of rational numbers,
e) $\{a\}$, j) the set $\mathbb{R} \setminus \mathbb{Q}$ of irrational numbers.

∘ **5.22** Show that $(-\infty, a]$, $[a, +\infty)$, $[a, b)$, $(a, b]$, and the set $\mathbb{Z}$ of whole numbers are Borel.

∗ **5.23** To construct the *Cantor set*, we remove the open interval $(\frac{1}{3}, \frac{2}{3})$ from the interval $[0, 1]$. Next, we divide each of the remaining two intervals into three equal subintervals and remove the middle ones, $(\frac{1}{9}, \frac{2}{9})$ and $(\frac{7}{9}, \frac{8}{9})$. In this manner, we carry on ad infinitum; see Figure 5.1, which shows the first five steps of the construction. The Cantor set is what remains after this infinite process.

Verify that the Cantor set is a Borel set.

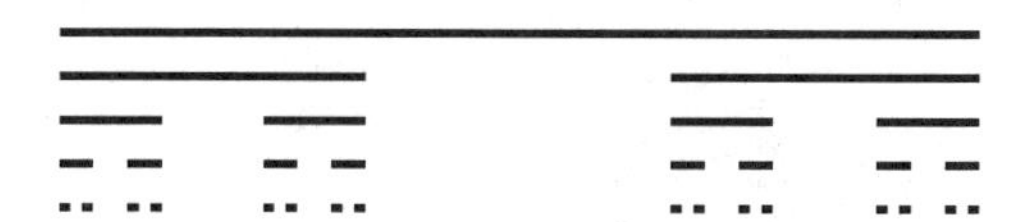

FIGURE 5.1. Construction of the Cantor set

∗ **5.24** Show that the $\sigma$-field $\mathcal{B}$ of Borel sets coincides with the $\sigma$-field generated by the family of open sets in $\mathbb{R}$.

∗ **5.25** Show that if $\mathcal{A}$ in Definition 5.3 is replaced by the family $\mathcal{C}$ of closed intervals, then the resulting $\sigma$-field is also $\mathcal{B}$.

∗ **5.26** Show that the inverse image of a Borel set under a continuous function $f : \mathbb{R} \to \mathbb{R}$ is Borel.

Approximating events by sequences of other events, a common task in probability theory, quite frequently involves the upper and lower limits of a sequence of sets as defined below.

▶ **Definition 5.4** We define the *upper limit* or *limit superior* of a sequence $A_1, A_2, \ldots \subset \Omega$ of sets by

$$\limsup_{n\to\infty} A_n = \bigcap_{n=1}^{\infty} \bigcup_{k=n}^{\infty} A_k$$

and the *lower limit* or *limit inferior* by

$$\liminf_{n\to\infty} A_n = \bigcup_{n=1}^{\infty} \bigcap_{k=n}^{\infty} A_k.$$

**5.27** Let $A_1, A_2, \ldots \in \mathcal{F}$, where $\mathcal{F}$ is a $\sigma$-field. Show that

$$\limsup_{n\to\infty} A_n \in \mathcal{F}, \qquad \liminf_{n\to\infty} A_n \in \mathcal{F}.$$

**5.28** Show that

$$\limsup_{n\to\infty} A_n = \{A_n \text{ occurs for infinitely many } n\},$$
$$\liminf_{n\to\infty} A_n = \{A_n \text{ occurs for all but finitely many } n\}.$$

**5.29** Find $\limsup_{n\to\infty} A_n$ and $\liminf_{n\to\infty} A_n$, where

$$A_n = \begin{cases} \left(\frac{1}{n}, \frac{2}{3} - \frac{1}{n}\right) & \text{if } n = 1, 3, 5, \ldots \\ \left(\frac{1}{3} - \frac{1}{n}, 1 + \frac{1}{n}\right) & \text{if } n = 2, 4, 6, \ldots . \end{cases}$$

**5.30** Show that $\liminf_{n\to\infty} A_n \subset \limsup_{n\to\infty} A_n$.

## 5.2 Hints

**5.1** It is sufficient to show condition (c) of Definition 3.1. You need to take two sets and extend them to a sequence of sets to apply condition (c) of Definition 5.1.

**5.2** Only condition (c) of Definition 5.1 needs to be verified. Employ the fact that there is a limited choice of countable sequences from a finite set.

**5.3** Here, $\mathcal{F}$ consists of all subsets of $\Omega$. So the condition $A \in \mathcal{F}$ reads: $A$ is a subset of $\Omega$.

**5.4** What happens when $\Omega$ is infinite, say $\Omega = \mathbb{N}$? Consider an infinite subset whose complement is also infinite. Can it be written as the union of a sequence of finite sets?

**5.5** Conditions (a) and (b) of Definition 5.1 follow immediately. To verify condition (c), take a sequence of sets in $\mathcal{F}$ and consider two cases: when all the sets are countable and when at least one of them has a countable complement.

**5.6** Express each of the given sets by countably many unions and complements of sets already known to be in $\mathcal{F}$.

**5.7** Write the intersection as the complement of the union of certain sets.

**5.8** Is it possible to write $\tilde{\Omega} = X^{-1}(A)$ for some $A \in \Omega$?
Is $\tilde{\Omega} \setminus X^{-1}(A)$ of the form $X^{-1}(B)$ for some $B \in \Omega$?
Is $\bigcup_{i=1}^{\infty} X^{-1}(A_i) = X^{-1}(B)$ for some $B \in \Omega$?

**5.9** Apply Problem 5.8 to $X : A \to \Omega$, the identity function on $A$.

**5.10** Does $X^{-1}(\Omega)$ belong to $\tilde{\mathcal{F}}$?
Does $X^{-1}(\Omega \setminus A)$ belong to $\tilde{\mathcal{F}}$ whenever $A$ belongs to $\mathcal{F}$?
Does $X^{-1}\left(\bigcup_{i=1}^{\infty} A_i\right)$ belong to $\tilde{\mathcal{F}}$ whenever $A_1, A_2, \ldots$ belong to $\mathcal{F}$?

**5.11** Observe that $X(A) \setminus X(B)$ is not always equal to $X(A \setminus B)$.
To construct an example, take an $\Omega$ having as few elements as possible (three elements will do).

**5.12** Consider a sequence of periodic sets whose periods tend to infinity. (Consult Hint 3.20 for the meaning of period.) Will the union of such sets be a periodic set?
For example, take

$$A_n = \bigcup_{l=-\infty}^{+\infty} (I_n + l i_n),$$

where $I_n = \{n\}$ and $i_n = 2^n$ for $n = 1, 2, \ldots$ . Is $\bigcup_{n=1}^{\infty} A_n$ a periodic set?

**5.13** This is, in fact, the same question as in Problem 3.16.

**5.14** See Problem 3.15.

**5.15** Consider (b) of Definition 5.1, for example. We shall show that it holds for $\bigcap_{j\in J}^{\infty} \mathcal{F}_j$. Suppose that $A \in \bigcap_{j\in J}^{\infty} \mathcal{F}_j$. Hence, $A \in \mathcal{F}_j$ for all $j$. Now, you can use (b), which is valid for each $\mathcal{F}_j$.

**5.16** Which of the given sets can be expressed by countably many unions, intersections, and complements of some sets in $\mathcal{A}$?
It becomes easier to answer this question if $\mathcal{A}$ can be replaced by a family $\tilde{\mathcal{A}}$ of pairwise disjoint non-empty sets such that $\mathcal{F}_{\mathcal{A}} = \mathcal{F}_{\tilde{\mathcal{A}}}$. Then, it suffices to verify whether or not any given set is a countable union of some sets from $\tilde{\mathcal{A}}$. Can you find such a family $\tilde{\mathcal{A}}$?

**5.17** The implications in both directions are rather obvious. Note that in the definition of $\mathcal{F}_{\mathcal{A}}$, we have the intersection of a family of $\sigma$-fields and $\mathcal{A}$ is one of them.

**5.18** Show that (a) implies that $\mathcal{F}_{\mathcal{A}} \subset \tilde{\mathcal{F}}$, whereas (b) implies the reverse inclusion.

**5.19** Verify conditions (a) and (b) of Problem 5.18.

**5.20** What is the relation between $\mathcal{C}$ and $\mathcal{F}_{\mathcal{D}}$?

**5.21** Employ the fact that $\mathcal{B}$ is a $\sigma$-field. Using the operations of complement, countable union, and countable intersection, represent each of the sets by means of open intervals and other sets already known to be in $\mathcal{B}$.

**5.22** See Problem 5.21.

**5.23** The Cantor set is constructed in stages. Is the set $C_n$ obtained at stage $n$ a Borel set? Express the Cantor set in terms of the sets $C_n$.

**5.24** Recall that a subset $U$ of the real line is open if for each $x \in U$, there is an $\varepsilon > 0$ such that $(x - \varepsilon, x + \varepsilon) \subset U$.

How can an open subset of the real line be represented by means of open intervals? Can it be done using countably many open intervals?

**5.25** Let $\mathcal{F}_{\mathcal{C}}$ denote the $\sigma$-field generated by closed intervals. Show that all open intervals are in $\mathcal{F}_{\mathcal{C}}$. From this, deduce that $\mathcal{B} \subset \mathcal{F}_{\mathcal{C}}$. The reverse inclusion can be obtained in a similar way.

**5.26** By Problem 5.10, the family of sets $\{A : f^{-1}(A) \text{ is Borel}\}$ is a $\sigma$-field on $\mathbb{R}$. Show that it contains all open sets, since $f$ is continuous. Next, use Problem 5.24.

**5.27** Each of the sets in question can be expressed by countable unions and intersections of $A_1, A_2, \ldots$ .

**5.28** How do you define the union and the intersection of a family of sets? Apply this to the definition of $\limsup A_n$ and $\liminf A_n$.

**5.29** Use Problem 5.28. Which points on the real line belong to infinitely many of the sets $A_n$? Which points belong to all but finitely many sets $A_n$?

**5.30** Use Problem 5.28.

## 5.3 Solutions

**5.1** Conditions (a) and (b) of Definition 5.1 are the same for fields and $\sigma$-fields. To verify (c), take any $A, B \in \mathcal{F}$. Put $A_1 = A$, $A_2 = B$, and $A_i = \emptyset$ for $i = 3, 4, 5, \ldots$ . Then, $A_i \in \mathcal{F}$ for all $i$, so by condition (c) of Definition 5.1

$$\bigcup_{i=1}^{\infty} A_i \in \mathcal{F}.$$

But this union is just $A \cup B$, which therefore belongs to $\mathcal{F}$.

**5.2** Let $\mathcal{F}$ be a finite field. Conditions (a) and (b) of Definition 5.1 are the same for fields and $\sigma$-fields. To verify (c), take a sequence

$A_1, A_2, \ldots \in \mathcal{F}$. Since $\mathcal{F}$ is finite, the sequence contains only finitely many distinct sets: $A_{i_1}, \ldots, A_{i_k}$, say. We know that

$$\bigcup_{l=1}^{k} A_{i_l} \in \mathcal{F},$$

since $\mathcal{F}$ is a field, and

$$\bigcup_{l=1}^{k} A_{i_l} = \bigcup_{i=1}^{\infty} A_i,$$

since the union will not change if we add any number of sets already included. This completes the proof.

**5.3** Here, $\mathcal{F}$ contains all subsets of $\Omega$, so it contains $\Omega$, $\Omega \setminus A$, and $\bigcup_{i=1}^{\infty} A_i$, no matter what $A$ and $A_i$ are.

**5.4** The answer is negative. For example, take $\Omega = \mathbb{N}$ and consider the set of even numbers $A \subset \Omega$. This set is the union of a sequence of finite sets,

$$A = \{2\} \cup \{4\} \cup \{6\} \cup \cdots,$$

but $A$ is not finite, nor is the complement $\Omega \setminus A$.

**5.5** The family in question is a $\sigma$-field.

Condition (a) of Definition 5.1 is satisfied because $\emptyset$ is a finite, and therefore a countable set.

Let $A \in \mathcal{F}$. Then, either $A$ is a countable set, in which case $\Omega \setminus A \in \mathcal{F}$ because it has a countable complement, or $A$ has a countable complement, which means that $\Omega \setminus A \in \mathcal{F}$ because $\Omega \setminus A$ is countable itself. Condition (b) of Definition 5.1 is therefore fulfilled.

Now, take any $A_1, A_2, \ldots \in \mathcal{F}$. If all the $A_i$ are countable, then $\bigcup_{i=1}^{\infty} A_i \in \mathcal{F}$ because a countable union of countable sets is countable, a well-known fact in set theory. If at least one of the $A_i$ has a countable complement, then so does $\bigcup_{i=1}^{\infty} A_i$ because $\Omega \setminus \bigcup_{i=1}^{\infty} A_i$ is contained in the complement $\Omega \setminus A_i$ for any $i = 1, 2, \ldots$ . This proves condition (c) of Definition 5.1.

**5.6** We express each of the given sets using complements and countable unions of sets that belong to $\mathcal{F}$ by assumption:

a) $\{0\} = [0, 1] \setminus \bigcup_{n=1}^{\infty} [\frac{1}{n+1}, \frac{1}{n}] \in \mathcal{F}$.

b) $\{\frac{1}{n} : n = 2, 3, \ldots\} = \bigcup_{n=2}^{\infty} \left( [\frac{1}{n+1}, \frac{1}{n}] \cap [\frac{1}{n}, \frac{1}{n-1}] \right) \in \mathcal{F}$.

c) $(\frac{1}{n}, 1] = \bigcup_{i=1}^{n-1} [\frac{1}{i+1}, \frac{1}{i}] \setminus [\frac{1}{n+1}, \frac{1}{n}] \in \mathcal{F}$.

d) $(0, \frac{1}{n}] = \bigcup_{i=n}^{\infty} [\frac{1}{i+1}, \frac{1}{i}] \in \mathcal{F}$.

**5.7** By de Morgan's law,

$$\bigcap_{i=1}^{\infty} A_i = \Omega \setminus \bigcup_{i=1}^{\infty} (\Omega \setminus A_i).$$

Now, $\Omega \in \mathcal{F}$ by (a) of Definition 5.1, all the $\Omega \setminus A_i$ are in $\mathcal{F}$ by (b), their union is in $\mathcal{F}$ by (c), and its complement is in $\mathcal{F}$ again by (b).

**5.8** First, $X^{-1}(\Omega) = \tilde{\Omega}$, so $\tilde{\Omega} \in \tilde{\mathcal{F}}$; that is, condition (a) of Definition 5.1 holds. Next, $X^{-1}(\Omega \setminus A) = \tilde{\Omega} \setminus X^{-1}(A)$, so (b) is satisfied. Finally, $X^{-1}(\bigcup_{i=1}^{\infty} A_i) = \bigcup_{i=1}^{\infty} X^{-1}(A_i)$, which shows (c).

**5.9** Let $X : A \to \Omega$ be defined by $X(\omega) = \omega$ for any $\omega \in A$. Then, $A \cap B = X^{-1}(B)$ for any $B \in \mathcal{F}$. This means that $\tilde{\mathcal{F}} = \{X^{-1}(B) : B \in \mathcal{F}\}$. By Problem 5.8, it follows that $\tilde{\mathcal{F}}$ is a $\sigma$-field on $A$.

**5.10** Let us verify the conditions of Definition 5.1.

Since $X^{-1}(\Omega) = \tilde{\Omega} \in \tilde{\mathcal{F}}$, it follows that $\Omega \in \mathcal{F}$. Condition (a) is satisfied.

Suppose that $A \in \mathcal{F}$. Then, $X^{-1}(A) \in \tilde{\mathcal{F}}$, which implies that $X^{-1}(\Omega \setminus A) = \tilde{\Omega} \setminus X^{-1}(A) \in \tilde{\mathcal{F}}$ because $\tilde{\mathcal{F}}$ is a $\sigma$-field. It follows that $\Omega \setminus A \in \mathcal{F}$, which proves (b).

Now, let $A_1, A_2, \ldots \in \mathcal{F}$. This means that $X^{-1}(A_1), X^{-1}(A_2), \ldots \in \tilde{\mathcal{F}}$, so $X^{-1}\left(\bigcup_{i=1}^{\infty} A_i\right) = \bigcup_{i=1}^{\infty} X^{-1}(A_i) \in \tilde{\mathcal{F}}$. It follows that $\bigcup_{i=1}^{\infty} A_i \in \mathcal{F}$, i.e., condition (c) holds.

**5.11** $\tilde{\mathcal{F}} = \{\emptyset, \{1\}, \{2,3\}, \{1,2,3\}\}$ is a $\sigma$-field on $\tilde{\Omega} = \{1,2,3\}$. Take $X : \{1,2,3\} \to \{1,2\}$ such that $X(1) = X(2) = 1$ and $X(3) = 2$.

Then, $\{X(A) : A \in \tilde{\mathcal{F}}\} = \{\emptyset, \{1\}, \{1,2\}\}$, which is not a $\sigma$-field, in fact not even a field because it has three elements (see Problem 3.26).

**5.12** Let us take

$$A_n = \bigcup_{l=-\infty}^{+\infty} (I_n + l i_n)$$

with $I_n = \{n\}$ and $i_n = 2^n$ for $n = 1, 2, \ldots$ . All the $A_n$ are periodic sets (see (3.1) in Problem 3.20). Is

$$A = \bigcup_{n=1}^{\infty} A_n$$

a periodic set or not?

Since $I_n = \{n\}$, it is clear that $n \in A$ for each $n = 1, 2, \ldots$ . But $0 \notin A$, for otherwise, $0 \in A_n$ for some $n$, implying that $n = 2^n$ for some $n$, which is impossible.

This means that $A$ cannot be periodic. If it were, then

$$A = \bigcup_{l=-\infty}^{+\infty} (I + li)$$

for some positive integer $i$ and some $I \subset \{1, \dots, i\}$. But every positive integer is in $A$, so $i \in A$. This would mean that $i \in I$, implying immediately that $0 \in A$ (take $l = -1$ in the above union), contrary to what has been shown above.

**5.13** Solution 3.16 also applies to $\sigma$-fields because every finite field is a $\sigma$-field (Problem 5.2).

**5.14** First, $\Omega \in \mathcal{F}_1$ and $\Omega \in \mathcal{F}_2$, so $\Omega \in \mathcal{F}_1 \cap \mathcal{F}_2$.

Let $A \in \mathcal{F}_1 \cap \mathcal{F}_2$. Then, $A \in \mathcal{F}_1$ and $A \in \mathcal{F}_2$; hence, $\Omega \setminus A \in \mathcal{F}_1$ and $\Omega \setminus A \in \mathcal{F}_2$. Thus, $\Omega \setminus A \in \mathcal{F}_1 \cap \mathcal{F}_2$.

Let $A_1, A_2, \ldots \in \mathcal{F}_1 \cap \mathcal{F}_2$. Then, $A_i \in \mathcal{F}_1$ and $A_i \in \mathcal{F}_2$ for all $i$; hence, $\bigcup_{i=1}^{\infty} A_i \in \mathcal{F}_1$ and $\bigcup_{i=1}^{\infty} A_i \in \mathcal{F}_2$. Thus, $\bigcup_{i=1}^{\infty} A_i \in \mathcal{F}_1 \cap \mathcal{F}_2$.

**5.15** First, $\Omega \in \mathcal{F}_j$ for all $j$, so $\Omega \in \bigcap_{j \in J} \mathcal{F}_j$.

Let $A \in \bigcap_{j \in J} \mathcal{F}_j$. Then, $A \in \mathcal{F}_j$, so $\Omega \setminus A \in \mathcal{F}_j$ for all $j \in J$; hence, $\Omega \setminus A \in \bigcap_{j \in J} \mathcal{F}_j$.

Let $A_1, A_2, \ldots \in \bigcap_{j \in J} \mathcal{F}_j$. Then, $A_i \in \mathcal{F}_j$ for all $i$ and $j$; hence, $\bigcup_{i=1}^{\infty} A_i \in \mathcal{F}_j$ for all $j$. Thus, $\bigcup_{i=1}^{\infty} A_i \in \bigcap_{j \in J} \mathcal{F}_j$.

**5.16** Consider the family $\tilde{\mathcal{A}} = \{0,1\} \cup \{[\frac{1}{2^{n+1}}, \frac{1}{2^n}) : n = 0, 1, 2, \ldots\}$, which consists of pairwise disjoint non-empty sets. We claim that $\mathcal{F}_{\mathcal{A}} = \mathcal{F}_{\tilde{\mathcal{A}}}$.

To verify the claim, it suffices to show that $\tilde{\mathcal{A}} \subset \mathcal{F}_{\mathcal{A}}$ and $\mathcal{A} \subset \mathcal{F}_{\tilde{\mathcal{A}}}$. This is because if $\tilde{\mathcal{A}} \subset \mathcal{F}_{\mathcal{A}}$, then $\mathcal{F}_{\tilde{\mathcal{A}}} \subset \mathcal{F}_{\mathcal{A}}$, and if $\mathcal{A} \subset \mathcal{F}_{\tilde{\mathcal{A}}}$, then $\mathcal{F}_{\mathcal{A}} \subset \mathcal{F}_{\tilde{\mathcal{A}}}$ by Definition 5.2.

But the inclusions $\tilde{\mathcal{A}} \subset \mathcal{F}_{\mathcal{A}}$ and $\mathcal{A} \subset \mathcal{F}_{\tilde{\mathcal{A}}}$ follow immediately because $\mathcal{A}$ and $\tilde{\mathcal{A}}$ differ only by the sets $\{0,1\}$ and $[0,1]$, for which

$$\{0,1\} = [0,1] \setminus \bigcup_{n=0}^{\infty} [\frac{1}{2^{n+1}}, \frac{1}{2^n}),$$

$$[0,1] = \{0,1\} \cup \bigcup_{n=0}^{\infty} [\frac{1}{2^{n+1}}, \frac{1}{2^n}).$$

Because $\tilde{\mathcal{A}}$ consists of pairwise disjoint non-empty sets, it follows that $\mathcal{F}_{\tilde{\mathcal{A}}}$ is the family of countable unions of sets from $\tilde{\mathcal{A}}$.

It remains to check which of the given sets can be written as a countable union of sets from $\tilde{\mathcal{A}}$. Clearly, $\{0,1\}, [\frac{1}{4}, 1), (0, \frac{1}{2}) \in \mathcal{F}_{\mathcal{A}}$ because they can be written in this way, whereas the remaining sets cannot. It follows that $\{0\}, \{1\}, \{\frac{1}{2}\}, \{\frac{1}{3}\}, (\frac{1}{4}, 1], [0, \frac{1}{2}] \notin \mathcal{F}_{\mathcal{A}}$.

**5.17** Assume that $\mathcal{F}_{\mathcal{A}} = \mathcal{A}$. By definition, $\mathcal{F}_{\mathcal{A}}$ is the intersection of a family of $\sigma$-fields. By Problem 5.15, it is a $\sigma$-field, so $\mathcal{A}$ is a $\sigma$-field.

Conversely, suppose that $\mathcal{A}$ is a $\sigma$-field. Since $\mathcal{A} \subset \mathcal{A}$, it follows that $\mathcal{A}$ is one of the $\sigma$-fields appearing in the family we intersect to obtain $\mathcal{F}_{\mathcal{A}}$. The intersection is contained in each of its terms, so $\mathcal{F}_{\mathcal{A}} \subset \mathcal{A}$. The reverse inclusion is obvious because $\mathcal{A}$ is contained in all the terms of the intersection.

**5.18** Because $\mathcal{F}_{\mathcal{A}}$ is a $\sigma$-field containing $\mathcal{A}$, it follows from (b) of Problem 5.18 that $\tilde{\mathcal{F}} \subset \mathcal{F}_{\mathcal{A}}$.

On the other hand, $\tilde{\mathcal{F}}$ is a $\sigma$-field containing $\mathcal{A}$ by condition (a). Therefore, it appears in the intersection defining $\mathcal{F}_{\mathcal{A}}$, which means that $\mathcal{F}_{\mathcal{A}} \subset \tilde{\mathcal{F}}$.

It follows that $\mathcal{F}_{\mathcal{A}} = \tilde{\mathcal{F}}$, completing the proof.

**5.19** Let $\tilde{\mathcal{F}}$ be the family consisting of all countable subsets of $\Omega$ and their complements. From Problem 5.5, we know that $\tilde{\mathcal{F}}$ is a $\sigma$-field.

Clearly, $\mathcal{A} \subset \tilde{\mathcal{F}}$ because every finite set is countable.

Every $\sigma$-field $\mathcal{F}$ that contains all finite subsets of $\Omega$ must also contain all countable subsets of $\Omega$. This is because every countable set $A \subset \Omega$ can be represented as the union of a sequence of finite subsets; namely, if $A = \{a_1, a_2, \dots\}$, then

$$A = \{a_1\} \cup \{a_2\} \cup \cdots .$$

By condition (b) of Definition 5.1, it follows that $\mathcal{F}$ must also contain the complement of every countable subset of $\Omega$. Therefore, if $\mathcal{A} \subset \mathcal{F}$, then $\tilde{\mathcal{F}} \subset \mathcal{F}$.

We have demonstrated that conditions (a) and (b) of Problem 5.18 are satisfied. It follows that $\tilde{\mathcal{F}} = \mathcal{F}_{\mathcal{A}}$, as required.

**5.20** Each $\sigma$-field in the intersection defining $\mathcal{F}_{\mathcal{D}}$ contains $\mathcal{D}$, so it also contains the smaller family $\mathcal{C}$. Hence, $\mathcal{C} \subset \mathcal{F}_{\mathcal{D}}$. By Problem 5.18, $\mathcal{F}_{\mathcal{C}}$ is the smallest $\sigma$-field containing $\mathcal{C}$. But $\mathcal{F}_{\mathcal{D}}$ is a $\sigma$-field containing $\mathcal{C}$; therefore, $\mathcal{F}_{\mathcal{C}} \subset \mathcal{F}_{\mathcal{D}}$.

**5.21**

a) $(a, b) \in \mathcal{A}$, so $(a, b) \in \mathcal{B}$.

b) $(a, +\infty) = \bigcup_{i \geqslant a} (a, i)$ and each $(a, i)$ is in $\mathcal{B}$, so condition (c) of Definition 5.1 implies the result because $\mathcal{B}$ is a $\sigma$-field.

c) $(-\infty, a) = \bigcup_{i \leqslant a} (i, a)$ and we use the above argument.

d) $[a, b] = \mathbb{R} \setminus [(-\infty, a) \cup (b, +\infty)]$ and we already know that the two infinite intervals are in $\mathcal{B}$, hence Problem 5.1 and condition (b) of Definition 5.1 yield the result.

e) $\{a\} = [a, a]$ is, in fact, an interval.

f) Any finite set is a finite union of one-element sets, which are in $\mathcal{B}$ by (e).

g) Any countable set is a countable union of one-element sets.

h) $\mathbb{N}$ is an example of a countable set, so the previous argument applies.

i) The set $\mathbb{Q}$ of rational numbers is in $\mathcal{B}$ as a countable set.

j) The set of $\mathbb{R}\setminus\mathbb{Q}$ of irrational numbers is the complement of $\mathbb{Q}$, which is in $\mathcal{B}$ by i).

**5.23** To construct the Cantor set, we take
$C_0 = [0, 1]$,
$C_1 = [0, \frac{1}{3}] \cup [\frac{2}{3}, 1]$,
$C_2 = [0, \frac{1}{9}] \cup [\frac{2}{9}, \frac{1}{3}] \cup [\frac{2}{3}, \frac{7}{9}] \cup [\frac{8}{9}, 1]$,
$\ldots$,
so at each stage we obtain the union of a finite number of intervals, which is Borel because each $\sigma$-field is a field (Problem 5.1). The Cantor set can be written as

$$C = \bigcap_{i=1}^{\infty} C_n,$$

which is in $\mathcal{B}$ by Problem 5.7.

**5.24** Denote the family of open sets by $\mathcal{U}$, and the family of open intervals by $\mathcal{A}$. By definition, $\mathcal{B} = \mathcal{F}_{\mathcal{A}}$. We have to prove that $\mathcal{F}_{\mathcal{A}} = \mathcal{F}_{\mathcal{U}}$. Each open interval is an open set, so $\mathcal{A} \subset \mathcal{U}$. Then, by Problem 5.20, $\mathcal{F}_{\mathcal{A}} \subset \mathcal{F}_{\mathcal{U}}$.

To prove the reverse inclusion, let $U$ be an open set. We claim that $U$ is a union of countably many intervals. To see this, consider all intervals contained in $U$ with rational ends. Their union is clearly contained in $U$. In fact, it is equal to $U$, since for any $x \in U$, we have $(x - \varepsilon, x + \varepsilon) \subset U$ for some $\varepsilon$, and we can find rational numbers $a$ and $b$ such that $x \in (a, b) \subset (x - \varepsilon, x + \varepsilon)$. This proves that $U \in \mathcal{F}_{\mathcal{A}}$ for each $U \subset \mathcal{U}$; that is, $\mathcal{U} \subset \mathcal{F}_{\mathcal{A}}$. Since $\mathcal{F}_{\mathcal{U}}$ is the smallest $\sigma$-field containing $\mathcal{U}$ (Problem 5.18), $\mathcal{F}_{\mathcal{U}} \subset \mathcal{F}_{\mathcal{A}}$.

**5.25** First, we show that all open intervals are in the $\sigma$-field $\mathcal{F}_{\mathcal{C}}$ generated by closed intervals. For any open interval $(a, b)$, we have $(a, b) = \bigcup_{n=1}^{\infty}[a + \frac{1}{n}, b - \frac{1}{n}]$. The intervals $[a + \frac{1}{n}, b - \frac{1}{n}]$ are in $\mathcal{F}_{\mathcal{C}}$, so the same is true for their countable union. The family $\mathcal{A}$ of open intervals is therefore contained in $\mathcal{F}_{\mathcal{C}}$. But $\mathcal{B}$ is the smallest $\sigma$-field containing $\mathcal{A}$, so $\mathcal{B} \subset \mathcal{F}_{\mathcal{C}}$.

Conversely, we know that every closed interval belongs to $\mathcal{B}$ (Problem 5.21), so $\mathcal{C} \subset \mathcal{B}$. Since $\mathcal{F}_{\mathcal{C}}$ is the smallest $\sigma$-field containing $\mathcal{C}$, it follows that $\mathcal{F}_{\mathcal{C}} \subset \mathcal{B}$.

**5.26** By Problem 5.10, the family $\mathcal{S} = \{A : f^{-1}(A) \in \mathcal{B}\}$ is a $\sigma$-field.

Since $f$ is continuous, the inverse image $f^{-1}(U)$ of any open set $U$ is open, so it belongs to the $\sigma$-field generated by open sets, which by Problem 5.24 is equal to $\mathcal{B}$. Therefore, $U \in \mathcal{S}$ for any open set $U \subset \mathbb{R}$.

It follows that $\mathcal{S}$ is a $\sigma$-field containing all open sets, so $\mathcal{B} \subset \mathcal{S}$ again by Problem 5.24. This means that $f^{-1}(A)$ is a Borel set for any $A \in \mathcal{B}$, as required.

**5.27** By condition (c) of Definition 5.1 and Problem 5.7, since $A_1, A_2, \ldots \in \mathcal{F}$, it follows that $\bigcup_{k=n}^{\infty} A_k \in \mathcal{F}$ and $\bigcap_{k=n}^{\infty} A_k \in \mathcal{F}$ for all $n = 1, 2, \ldots$ , which, in turn, implies that $\bigcap_{n=1}^{\infty} \bigcup_{k=n}^{\infty} A_k \in \mathcal{F}$ and $\bigcup_{n=1}^{\infty} \bigcap_{k=n}^{\infty} A_k \in \mathcal{F}$. This completes the proof.

**5.28** By Definition 5.4,

$$
\begin{aligned}
\omega \in \limsup_{n\to\infty} A_n &\iff \omega \in \bigcap_{n=1}^{\infty} \bigcup_{k=n}^{\infty} A_k \\
&\iff \forall n \, \exists k \geqslant n : \omega \in A_k \\
&\iff \omega \in A_n \text{ for infinitely many } n, \\
\omega \in \liminf_{n\to\infty} A_n &\iff \omega \in \bigcup_{n=1}^{\infty} \bigcap_{k=n}^{\infty} A_k \\
&\iff \exists n \, \forall k \geqslant n : \omega \in A_k \\
&\iff \omega \in A_n \text{ for all but finitely many } n.
\end{aligned}
$$

**5.29** $\limsup_{n\to\infty} A_n = (0, 1]$, $\liminf_{n\to\infty} A_n = [\frac{1}{3}, \frac{2}{3})$.

**5.30** If $A_n$ occurs for all but finitely many $n$, then it occurs for infinitely many $n$. By Problem 5.28, this proves the inclusion.

# 6

# Countably Additive Probability

## 6.1 Theory and Problems

We shall extend the notion of probability measure from fields to $\sigma$-fields. The main point here is that we shall assume the probability measure $P$ to be *countably additive* rather than just finitely additive. This allows us to tackle situations in which the number of possible outcomes of a random experiment is infinite.

**Example 6.1** A coin is tossed repeatedly until it lands heads up. The number of tails before the first head appears can be $0, 1, 2, \ldots$. This can be modeled by the probability space $\Omega = \{0, 1, 2, \ldots\}$, where $P(\{n\}) = \frac{1}{2^{n+1}}$ is the probability of obtaining $n$ tails before the first head, as in the solution to Problem 1.9.

Suppose you have made a bet that $n$ will be even. How likely is it to happen? Let $\mathcal{F}$ be the $\sigma$-field consisting of all subsets of $\Omega$, so $\{n \text{ is even}\} \in \mathcal{F}$ is an event. Because

$$\{n \text{ is even}\} = \{0\} \cup \{2\} \cup \{4\} \cup \cdots ,$$

it is natural to take

$$P(\{n \text{ is even}\}) = \frac{1}{2} + \frac{1}{2^3} + \frac{1}{2^5} + \cdots = \frac{2}{3} .$$

The basic idea is the same as in the case of finitely additive probability: Condition (b) of Definition 6.1 means that to measure a set, we can decompose it into a sequence of disjoint pieces, measure each piece separately, and then add up the results. However, using countably many pieces, we will be able to measure a much greater variety of events than was possible with only finitely many pieces. This is rather like measuring the area of a circle by decomposing it into infinitely many rectangles.

▶ **Definition 6.1** We call a function $P : \mathcal{F} \to [0,1]$ a *probability measure* if

(a) $P(\Omega) = 1$;

(b) (*countable additivity*) for any sequence $A_1, A_2, \ldots \in \mathcal{F}$ of pairwise disjoint sets (that is, $A_i \cap A_j = \emptyset$ if $i \neq j$)

$$P\left(\bigcup_{i=1}^{\infty} A_i\right) = \sum_{i=1}^{\infty} P(A_i).$$

**6.1** Show that if $P$ is countably additive, then it is also finitely additive; that is, $P(A \cup B) = P(A) + P(B)$ for any $A, B \in \mathcal{F}$ such that $A \cap B = \emptyset$.

**6.2** Let $\mathcal{F}$ be a finite field. Show that if $P$ is a finitely additive probability measure on $\mathcal{F}$, then it is also countably additive.

In the following, we have a series of problems similar to Problems 4.7 and 4.8 concerned with extending $P$ to a probability measure.

**6.3** Let $\Omega = \mathbb{N}$ and let $\mathcal{F}$ be the family of all subsets of $\Omega$. Can $P(\{i\}) = \frac{1}{i}$, $i = 1, 2, \ldots$, be extended to a probability measure on $\mathcal{F}$?

**6.4** Let $\Omega = \mathbb{N}$, and let $\mathcal{F}$ be the family of all subsets of $\Omega$. Put $P(\{i\}) = \alpha_i$, $i = 1, 2, \ldots$ . Extend $P$ to a probability measure defined on $\mathcal{F}$. What conditions have to be imposed on the numbers $\alpha_i$? Can they all be chosen the same?

**6.5** Let $\Omega = \mathbb{N}$ and let $\mathcal{F}$ be the family of all subsets of $\Omega$. Can $P(\{i\}) = 2^{-i}$, $i = 1, 2, \ldots$, be extended to a probability measure on $\mathcal{F}$?

**6.6** Let $\Omega = \mathbb{N}$ and let $\mathcal{F}$ be the family of all subsets of $\Omega$. Can you find a value of $c$ such that $P(\{i\}) = \frac{c}{i!}$, $i = 1, 2, \ldots$, can be extended to a probability measure on $\mathcal{F}$?

* **6.7** Let $\Omega = \mathbb{N}$ and let $\mathcal{F}$ be the family of all subsets of $\Omega$. Is

$$P(A) = \liminf_{n \to \infty} \frac{\#(A \cap \{1, \ldots, n\})}{n}$$

a probability measure on $\mathcal{F}$?

Let $A_1, A_2, \ldots \in \mathcal{F}$ be a sequence of events. Show that:

▶ **6.8** If $A_1 \subset A_2 \subset A_3 \subset \cdots$ (this is called an *expanding* or *increasing* sequence of sets), then

$$P\left(\bigcup_{k=1}^{\infty} A_k\right) = \lim_{k\to\infty} P(A_k).$$

▶ **6.9** If $A_1 \supset A_2 \supset A_3 \supset \cdots$ (this is called a *contracting* or *decreasing* sequence of sets), then

$$P\left(\bigcap_{k=1}^{\infty} A_k\right) = \lim_{k\to\infty} P(A_k).$$

▶ **6.10** $P\left(\bigcup_{n=1}^{\infty} A_n\right) \leqslant \sum_{i=1}^{\infty} P(A_n);$

**6.11** If $P(A_n) = 0$ for $n = 1, 2, \ldots$, then $P\left(\bigcup_{n=1}^{\infty} A_n\right) = 0$.

**6.12** If $P(A_n) = 1$ for $n = 1, 2, \ldots$, then $P\left(\bigcap_{n=1}^{\infty} A_n\right) = 1$.

**6.13** If $P(A_n \triangle A) \to 0$ as $n \to \infty$, then $P(A_n) \to P(A)$ as $n \to \infty$.

**6.14** Prove Theorem 6.1. (It is called the *first Borel-Cantelli lemma.*)

▶ **Theorem 6.1** *If* $\sum_{n=1}^{\infty} P(A_n) < \infty$, *then* $P(\limsup_{n\to\infty} A_n) = 0$.

∗ **6.15** Verify that

$$P(\liminf_{n\to\infty} A_n) \leqslant \liminf_{n\to\infty} P(A_n) \leqslant \limsup_{n\to\infty} P(A_n) \leqslant P(\limsup_{n\to\infty} A_n).$$

We wish to introduce some interesting examples of probability measures defined on $\mathbb{R}$. To do this, we need the concept of *Lebesgue measure.* Lebesgue measure on $\mathbb{R}$ is an extension of the notion of length. Since the "length" of $\mathbb{R}$, whatever it might be, is certainly infinite, we are forced to allow the values of Lebesgue measure to be in $[0, \infty]$, with the infinity symbol $\infty$ formally added to the set of all non-negative real numbers.

▶ **Definition 6.2** A function $m$ on the $\sigma$-field $\mathcal{B}$ of Borel subsets of $\mathbb{R}$,

$$m : \mathcal{B} \to [0, \infty],$$

mapping each finite interval into its length (i.e., $m([a, b]) = m((a, b)) = m([a, b)) = m((a, b]) = b - a$) such that

$$m\left(\bigcup_{i=1}^{\infty} A_i\right) = \sum_{i=1}^{\infty} m(A_i)$$

for any sequence of pairwise disjoint Borel sets $A_1, A_2, \ldots$ is called *Lebesgue measure.*

Hence, $P = m$ is a probability measure on $\Omega = [0,1]$ with the $\sigma$-field $\mathcal{F} = \{A \subset [0,1] : A \in \mathcal{B}\}$.

A precise construction of this measure is beyond the scope of this book. We just assume that Lebesgue measure exists. The underlying idea is simple: We define $m$ on intervals and then extend it to Borel sets. However, the precise construction of $m$ presents some difficulties. Fortunately, all we need for our purposes is to compute $m(A)$ for relatively simple sets $A$ obtained from intervals by countable unions and intersections.

*Remark* Lebesgue measure can, in fact, be extended to a $\sigma$-field bigger than $\mathcal{B}$, but cannot be defined for all subsets of $\mathbb{R}$.

**6.16** Find $m([\frac{1}{2},\frac{2}{3}))$, $m((\frac{1}{5},\frac{1}{3}])$, $m(\{1\})$, $m(\{\frac{1}{n} : n = 1,2,\dots\})$, and $m(\mathbb{N})$.

○ **6.17** Find $m([\frac{1}{4},\frac{1}{2}])$, $m(\{0,\frac{1}{2},1\})$, and $m(\mathbb{Q} \cap [0,1])$ ($\mathbb{Q}$ denotes the set of all rational numbers).

**6.18** Find $m(\{x : |x-n| < \frac{1}{2^n}$ for some $n \in \mathbb{N}\})$.

○ **6.19** Find $m(\bigcup_{n=1}^{\infty}(\frac{1}{2^{n+1}},\frac{1}{2^n}])$.

* **6.20** Find $m(C)$, where $C$ is the Cantor set (see Problem 5.23 for the definition of the Cantor set).

Other interesting examples of probability measures can be constructed with the aid of the integral with respect to Lebesgue measure. Here, we shall confine ourselves to integrating non-negative *Borel functions* only, i.e., functions $f : \mathbb{R} \to [0,\infty)$ such that $\{x : f(x) < a\}$ is a Borel set for every $a \in \mathbb{R}$. In particular, each non-negative piecewise continuous function belongs to this class.

▶ **Definition 6.3** The *Lebesgue integral* is a mapping

$$f \mapsto \int f \, \mathrm{d}m \in [0,\infty]$$

defined for each non-negative Borel function $f : \mathbb{R} \to [0,\infty)$ such that

(a) $\int \mathbb{I}_A \, \mathrm{d}m = m(A)$,

(b) (*linearity*) $\int (f+g) \, \mathrm{d}m = \int f \, \mathrm{d}m + \int g \, \mathrm{d}m$ and $\int cf \, \mathrm{d}m = c \int f \, \mathrm{d}m$,

(c) (*monotone convergence*) if $f_n \nearrow f$, then $\int f_n \, \mathrm{d}m \nearrow \int f \, \mathrm{d}m$

for any Borel set $A \subset \mathbb{R}$, any $c \geqslant 0$, and any non-negative Borel functions $f, g$ and $f_1, f_2, \dots$ . (Here, $f_n \nearrow f$ means that $f_n(x) \nearrow f(x)$ for each $x \in \mathbb{R}$.)

We also define

$$\int_A f \, \mathrm{d}m = \int f \mathbb{I}_A \, \mathrm{d}m$$

for any Borel set $A \subset \mathbb{R}$.

*Remark* The Lebesgue integral can, in fact, be defined for a class of functions larger than Borel functions, but not for all real-valued functions.

As with the Lebesgue measure, the precise construction of the Lebesgue integral is beyond the scope of this book. Suffice it to say that for any non-negative piecewise continuous function $f$, it reduces to the Riemann integral,

$$\int_{[a,b]} f\,\mathrm{d}m = \int_a^b f(x)\,\mathrm{d}x = F(b) - F(a),$$

where $F$ is the antiderivative of $f$, i.e., $F' = f$. For $(-\infty, +\infty)$, $(-\infty, a]$, or $[a, +\infty)$, we need to compute the improper Riemann integral; for example,

$$\int_{[a,+\infty)} f\,\mathrm{d}m = \int_a^\infty f(x)\,\mathrm{d}x = \lim_{b\to\infty} F(b) - F(a).$$

This is all we shall need in this course.

Because our interest lies in probability measures, the following notion is relevant.

▶ **Definition 6.4** We call a non-negative Borel function $f : \mathbb{R} \to [0, \infty)$ a *density* if

$$\int f\,\mathrm{d}m = 1.$$

Are the following functions $f$ densities? (Choose the constant $c$ if necessary.)

**6.21** $f(x) = c\mathbb{I}_{[1,3]}(x)$.

**6.22** $f(x) = c\mathbb{I}_{[a,b]}(x)$.

**6.23** $f(x) = cx^n$ for $x \in (0, 1)$ and zero otherwise, where $n \in \mathbb{N}$.

**6.24** $f(x) = ce^{-\lambda x}$ for $x > 0$ and zero otherwise.

∘ **6.25** $f(x) = cx(1 - x)$ for $x \in [0, 1]$ and zero otherwise.

∘ **6.26** $f(x) = c(x - [x])$ for $x \in [0, 2]$ and zero otherwise ($[x]$ denotes the largest integer not greater than $x$).

∘ **6.27** $f(x) = x^2\mathbb{I}_{[0,c]}(x)$.

The notion of density allows us to specify an important class of probability measures on the real line.

▶ **Theorem 6.2** *If $f$ is a density, then*

$$P(A) = \int_A f\, dm$$

*is a probability measure on Borel subsets $A$ of $\mathbb{R}$.*

**6.28** Prove Theorem 6.2.

▶ **Definition 6.5** A probability measure $P$ on Borel subsets $A$ of $\mathbb{R}$ is said to be *absolutely continuous* if there exists a density $f$ such that

$$P(A) = \int_A f\, dm.$$

Find $P(A)$ if $P$ is an absolutely continuous probability measure with density $f$, where:

**6.29** $f(x) = \frac{1}{2}\mathbb{I}_{[1,3]}(x)$, $A = [0, 2)$;

**6.30** $f(x) = x - [x]$ for $x \in [0, 2]$ and zero otherwise, $A = [\frac{1}{2}, 2]$ ($[x]$ denotes the largest integer not greater than $x$);

**6.31** $f(x) = e^{-x}$ for $x \geqslant 0$ and zero otherwise, $A = \mathbb{N}$;

∘ **6.32** $f(x) = \frac{1}{5}\mathbb{I}_{[-2,3]}(x)$, $A = (0, 1)$;

∘ **6.33** $f(x) = \frac{1}{2}x^{-\frac{1}{2}}$ for $x \in (0, 1)$ and zero otherwise, $A = (0, \frac{1}{2})$.

In an arbitrary set $\Omega$, we can introduce probability measures which are, in fact, equivalent to measures defined on certain finite or countable subsets. We begin with a simple but famous case.

▶ **Definition 6.6** Let us take any $\Omega$ and suppose that $\mathcal{F}$ is the family of all subsets of $\Omega$. We fix an element $\omega \in \Omega$ and put

$$\delta_\omega(A) = \begin{cases} 1 & \text{if } \omega \in A \\ 0 & \text{otherwise.} \end{cases}$$

We call $\delta_\omega$ the *Dirac measure* concentrated at $\omega$.

**6.34** Show that $\delta_\omega$ is a probability measure. What is the largest set of measure 0? What is the smallest set of measure 1?

**6.35** Let $\omega_1, \omega_2 \in \Omega$. Show that $P(A) = \frac{1}{2}\delta_{\omega_1}(A) + \frac{1}{2}\delta_{\omega_2}(A)$ is a probability measure. Which sets have measure 0?

**6.36** What are the conditions on $\alpha_k$ for $P = \sum_{k=1}^{\infty} \alpha_k \delta_{\omega_k}$ to be a probability measure? Which sets have measure 0?

▶ **Definition 6.7** Probability measures of the form

$$P(A) = \sum_{k=1}^{\infty} \alpha_k \delta_{\omega_k}(A),$$

where $\alpha_k \geqslant 0$ for all $k = 1, 2, \ldots$ and $\sum_{k=1}^{\infty} \alpha_k = 1$, are called *discrete.*

We close this section with examples of measures on $\mathbb{R}$ which are neither discrete nor absolutely continuous. First, a general fact.

**6.37** If $P_1$ and $P_2$ are probability measures, then $P(A) = \alpha_1 P_1(A) + \alpha_2 P_2(A)$ is also a probability measure provided that $\alpha_1$ and $\alpha_2$ are non-negative and $\alpha_1 + \alpha_2 = 1$.

$*$ **6.38** If $P_k$, $k = 1, 2, \ldots$, are probability measures, $\alpha_k \geqslant 0$ for all $k$, and $\sum_{k=1}^{\infty} \alpha_k = 1$, then $P(A) = \sum_{k=1}^{\infty} \alpha_k P_k(A)$ is a probability measure.

**6.39** Define $P$ by $P(A) = \frac{1}{3}\delta_0(A) + \frac{2}{3}P_2(A)$, where $P_2$ has density $f(x) = \frac{1}{2}\mathbb{I}_{[1,3]}(x)$. Compute $P([0, 2])$.

$\circ$ **6.40** Define $P$ by $P(A) = \frac{1}{8}\delta_1(A) + \frac{3}{8}\delta_2(A) + \frac{1}{2}P_3(A)$, where $P_3$ is given by the density $f(x) = \frac{1}{2}\mathbb{I}_{[1,3]}(x)$. Compute $P([2, 3])$.

$\circ$ **6.41** Define $P$ by $P(A) = \frac{1}{4}\delta_0(A) + \frac{1}{2}\delta_2(A) + \frac{1}{4}P_3(A)$, where $P_3$ has density $f(x) = e^{-x}\mathbb{I}_{[0,+\infty)}(x)$. Compute $P([0, 1])$.

We conclude with a brief observation that it is impossible to go beyond countable additivity in the definition of probability measure.

$*$ **6.42** Show that if $X$ is an uncountable set of positive real numbers, then there is a sequence $x_1, x_2, \ldots \in X$ such that $\sum_{i=1}^{\infty} x_i = \infty$.

The assertion in Problem 6.42 means that countable additivity is all that can be required of a probability measure because no uncountable family of disjoint sets $A$ with $P(A) > 0$ can exist anyway, or otherwise $P(\Omega)$ would have to be $\infty$ rather than 1.

## 6.2 Hints

**6.1** Of course, we need condition (b) of Definition 6.1, but it involves a countable sequence of events. Try to extend the finite sequence of events $A$ and $B$ to a suitable infinite sequence without affecting either side of the equality to be proven.

**6.2** If $\mathcal{F}$ is finite, then a sequence of pairwise disjoint sets from $\mathcal{F}$ has all except finitely many entries empty.

**6.3** A necessary condition is $P(\Omega) = 1$.

**6.4** The numbers $\alpha_i$ are some of the values of $P$. First, the question: What are the admissible values of $P$? This gives you some restrictions on the $\alpha_i$. Next, take into account that $P(\Omega) = 1$.

To extend $P$, give a formula for $P(A)$ where $A \subset \Omega$ has $m$ elements, say $A = \{k_1, \ldots, k_m\}$. For example, consider $A = \{1, 2\}$. Try to make the notation as simple as possible. Then, extend your formula to infinite sets $A$.

**6.5** See Problem 6.4.

**6.6** See Problem 6.4.

**6.7** First, check that $P(\Omega) = 1$. Next, find $P(A)$ for one-element sets $A = \{k\}$. Is the latter consistent with the former?

**6.8** We know how to deal with the union of pairwise disjoint sets. Draw a picture of a few sets $A_i$, each included in the next one. How does one define another family of sets, this time pairwise disjoint, such that the union is the same?

**6.9** De Morgan's laws allow us to express intersections of sets in terms of unions or vice versa. Consider the set $\Omega \setminus \bigcap_{k=1}^{\infty} A_k$ and try to apply Problem 6.8.

**6.10** Apply Problem 6.8 to the sequence $B_n = A_1 \cup \cdots \cup A_n$ and then use Problem 4.17.

**6.11** Use Problem 6.10.

**6.12** Use Problem 6.11 and de Morgan's law.

**6.13** Use Problem 4.24.

**6.14** Try to show that $P\left(\bigcup_{n=k}^{\infty} A_n\right)$ can be arbitrarily small if $k$ is large enough. To this end, use Problem 6.10. Can $P(\limsup_{n\to\infty} A_n)$ be larger than any of these probabilities?

**6.15** To use Problems 6.8 and 6.9, you need expanding and contracting sequences. Consider $B_n = \bigcap_{k=n}^{\infty} A_k$ and $C_n = \bigcup_{k=n}^{\infty} A_k$. How are these related to $\liminf_{n\to\infty} A_n$ and $\limsup_{n\to\infty} A_n$?

**6.16** The value of $m$ on intervals is given in Definition 6.2, so the first two questions are straightforward. Note that any one-element set is an interval. For the last two sets, use countable additivity ($m$, as assumed, satisfies condition (b) of Definition 6.1).

**6.17** See Problem 6.16.

**6.18** Solve the inequality $|x - n| < \frac{1}{2^n}$ for $x$. The solution will depend on $n$. Because the phrase "for some $n$" has been used, you must consider the union.

**6.19** $\bigcup_{n=1}^{\infty}(\frac{1}{2^{n+1}}, \frac{1}{2^n}]$ is a union of pairwise disjoint intervals. Find the measure of each of the intervals and use countable additivity.

**6.20** Use Problem 6.9.

**6.21** Here, we have the integral of a very simple function:

$$f(x) = \begin{cases} 0 & \text{for } x < 1 \\ c & \text{for } x \in [1, 3] \\ 0 & \text{for } x > 3. \end{cases}$$

**6.22** See Problem 6.21.

**6.23** All you need to know is how to compute the integral of $x^n$ from 0 to 1.

**6.24** See Problem 6.23. Some calculus is required, namely improper integrals.

**6.25** See Problem 6.23.

**6.26** See Problem 6.23. Split the interval into $[0, 1]$ and $[1, 2]$ and find the explicit form of $f$ in each of these.

**6.27** See Problem 6.23. This time you need to integrate from 0 to $c$.

**6.28** $P(\mathbb{R})$ is equal to the integral of $f$ over $\mathbb{R}$. To prove countable additivity, use monotone convergence (condition (c) of Definition 6.3).

**6.29** We have to integrate $f$ over $[0, 2)$. To do this, split the interval $[0, 2)$ into $[0, 1)$ and $[1, 2)$.

**6.30** Split $A$ into two intervals, $[\frac{1}{2}, 1]$ and $(1, 2]$, to transform $[x]$.

**6.31** $P$ is a probability measure, so you can use condition (b) of Definition 6.1.

**6.32** Compute the integral of $f$ over $(0, 1)$; see Problem 6.29.

**6.33** Compute the integral of $f$ over $(0, \frac{1}{2})$. It is going to be an improper integral, so a limit procedure will be required.

**6.34** Are the values of $\delta_\omega$ in $[0,1]$? Condition (a) of Definition 6.1 should present no difficulty. To verify (b), suppose that $\omega \in \bigcup_{i=1}^\infty A_i$ and compute both sides of the equality. Then, suppose that $\omega \notin \bigcup_{i=1}^\infty A_i$ and compute both sides of (b) again.

**6.35** See Problem 6.34, but here you have to consider more cases. Which of the $\omega_i$ belong to $\bigcup_{i=1}^\infty A_i$: one, both, or neither?

Alternatively, you can use Problem 6.34 for each of the two terms in $P = \frac{1}{2}\delta_{\omega_1} + \frac{1}{2}\delta_{\omega_2}$.

**6.36** What is $P(\Omega)$ here? Do not forget the usual assumption about the admissible values of $P$.

**6.37** When checking the conditions of Definition 6.1 for $P$, employ the corresponding ones for $P_i$. For example, $P(\Omega) = \alpha_1 P_1(\Omega) + \alpha_2 P_2(\Omega) = \alpha_1 \times 1 + \alpha_2 \times 1 = 1$. Now it is your turn.

**6.38** See Problem 6.37 for a particular case ($\alpha_k = 0$ for $k > 2$). A similar argument can be applied here.

**6.39** You need to know how to compute the Dirac measure of a set (see Definition 6.6) and the integral of an indicator function (see Problem 6.29).

**6.40** See Problem 6.39.

**6.41** See Problem 6.39.

**6.42** The set $\left\{x \in X : x > \frac{1}{n}\right\}$ must be infinite for some $n = 1, 2, \ldots$ if $X$ is uncountable. Once this is established, the statement follows easily.

## 6.3 Solutions

**6.1** Put $A_1 = A$, $A_2 = B$, and $A_i = \emptyset$ for $i > 2$, and use condition (b) of Definition 6.1:

$$\begin{aligned}
P(A \cup B) &= P\left(\bigcup_{i=1}^\infty A_i\right) && \text{by the definition of } A_i \\
&= \sum_{i=1}^\infty P(A_i) && \text{by countable additivity} \\
&= P(A) + P(B) && \text{since } P(A_i) = 0 \text{ for } i > 2.
\end{aligned}$$

**6.2** Suppose that $A_i \in \mathcal{F}$ and $A_i \cap A_j = \emptyset$ for $i \neq j$. Because $\mathcal{F}$ is finite, we can find an $n$ such that $A_i = \emptyset$ for $i > n$, so $P(A_i) = 0$ for those $i$. Then,

$$\bigcup_{i=1}^{\infty} A_i = \bigcup_{i=1}^{n} A_i$$

and we can apply Problem 4.14 to get

$$\begin{aligned} P\left(\bigcup_{i=1}^{\infty} A_i\right) &= P\left(\bigcup_{i=1}^{n} A_i\right) && \text{since } A_i = \emptyset \text{ for } i > n \\ &= \sum_{i=1}^{n} P(A_i) && \text{by finite additivity} \\ &= \sum_{i=1}^{\infty} P(A_i) && \text{since } P(A_i) = 0 \text{ for } i > n. \end{aligned}$$

**6.3** Since $\sum_{i=1}^{\infty} \frac{1}{i} = \infty$, the extension is impossible: The measure of $\Omega = \mathbb{N}$ would have to be infinite.

**6.4** Since the values of $P$ must be non-negative, we assume that $0 \leqslant \alpha_i$ for all $i$. To have $P(\Omega) = 1$, we must assume that

$$\sum_{i=1}^{\infty} \alpha_i = 1.$$

The convergence of the series implies that $\alpha_i \to 0$ as $i \to \infty$, so the numbers $\alpha_i$ cannot be the same.

For any $A \subset \Omega$, we put

$$P(A) = \sum_{i \in A} P(\{i\}) = \sum_{i \in A} \alpha_i.$$

Proof of countable additivity (condition (b) of Definition 6.1): Let $A_1, A_2, \ldots \subset \Omega$ be disjoint sets. Then,

$$\begin{aligned} P\left(\bigcup_{k=1}^{\infty} A_k\right) &= \sum_{i \in \bigcup_{k=1}^{\infty} A_k} \alpha_i \\ &= \sum_{i \in A_1} \alpha_i + \sum_{i \in A_2} \alpha_i + \cdots \\ &= P(A_1) + P(A_2) + \cdots . \end{aligned}$$

Because $\alpha_i \geqslant 0$ and $\sum_{i=1}^{\infty} \alpha_i < \infty$, the above series are convergent, their sums being independent of the order of terms.

Note that $i$ and $k$ play different roles: $i$ denotes an element of $\Omega$ and $k$ is the index (consecutive number) of a set in the sequence $A_1, A_2, \ldots$ . It is a coincidence ($\Omega = \mathbb{N}$) that $i$ and $k$ run through the same set.

**6.5** Since $\frac{1}{2^i} \geqslant 0$ and $\sum_{i=1}^{\infty} \frac{1}{2^i} = 1$, it follows from Problem 6.4 that $P(\{i\}) = 2^{-i}$ can be extended to a probability measure on $\mathcal{F}$.

**6.6** Since $\sum_{i=1}^{\infty} \frac{1}{i!} = e$, by putting $c = \frac{1}{e}$ we get $\frac{c}{i!} \geqslant 0$ and $\sum_{i=1}^{\infty} P(\{i\}) = \sum_{i=1}^{\infty} \frac{c}{i!} = 1$; therefore 6.4 can be applied to obtain the extension.

**6.7** To find $P(\mathbb{N})$, observe that $\mathbb{N} \cap \{1, \ldots, n\} = \{1, \ldots, n\}$ has $n$ elements and $P(\mathbb{N}) = \liminf_{n\to\infty} \frac{n}{n} = 1$.

For $A = \{k\}$, the set $A \cap \{1, \ldots, n\}$ is empty if $n < k$ and has just one element $k$ if $n \geqslant k$. Thus, the sequence in the formula for $P(\{k\})$ begins with $k - 1$ zeros followed by $\frac{1}{k}, \frac{1}{k+1}, \ldots$ , which converges to 0. Therefore, $P(\{k\}) = 0$. This gives a contradiction, since $P(\mathbb{N}) = \sum_{k=1}^{\infty} P(\{k\}) = 0$ by countable additivity.

**6.8** Put $B_k = A_{k+1} \setminus A_k$. The sets $B_k$ are pairwise disjoint, so condition (b) of Definition 6.1 can be applied. Moreover, we have

$$\bigcup_{k=1}^{\infty} B_k = \bigcup_{k=1}^{\infty} A_k \quad \text{and} \quad \bigcup_{k=1}^{n} B_k = A_n.$$

It follows that

$$\begin{aligned}
P\left(\bigcup_{k=1}^{\infty} A_k\right) &= P\left(\bigcup_{k=1}^{\infty} B_k\right) \\
&= \sum_{k=1}^{\infty} P(B_k) && \text{by countable additivity} \\
&= \lim_{n\to\infty} \sum_{k=1}^{n} P(B_k) && \text{by the definition of the sum of a series} \\
&= \lim_{n\to\infty} P\left(\bigcup_{k=1}^{n} B_k\right) && \text{by additivity again} \\
&= \lim_{n\to\infty} P(A_n) && \text{since } \textstyle\bigcup_{k=1}^{n} B_k = A_n.
\end{aligned}$$

**6.9** The sets $\Omega \setminus A_k$ form an expanding sequence, so Problem 6.8 can be used:

$$\begin{aligned} 1 - P\left(\bigcap_{k=1}^{\infty} A_k\right) &= P\left(\Omega \setminus \bigcap_{k=1}^{\infty} A_k\right) && \text{by Problem 4.11} \\ &= P\left(\bigcup_{k=1}^{\infty} (\Omega \setminus A_k)\right) && \text{by de Morgan's law} \\ &= \lim_{k\to\infty} P(\Omega \setminus A_k) && \text{by Problem 6.8} \\ &= 1 - \lim_{k\to\infty} P(A_k) && \text{again by Problem 4.11.} \end{aligned}$$

**6.10** Put $B_n = A_1 \cup \cdots \cup A_n$. Then,

$$\begin{aligned} P\left(\bigcup_{n=1}^{\infty} A_n\right) &= P\left(\bigcup_{n=1}^{\infty} B_n\right) && \text{since } \bigcup_{n=1}^{\infty} A_n = \bigcup_{n=1}^{\infty} B_n \\ &= \lim_{n\to\infty} P(B_n) && \text{by Problem 6.8, since } B_1 \subset B_2 \subset \cdots \\ &\leqslant \lim_{n\to\infty} \sum_{k=1}^{n} P(A_k) && \text{by Problem 4.17} \\ &= \sum_{n=1}^{\infty} P(A_n) && \text{by the definition of the sum of a series.} \end{aligned}$$

**6.11** By Problem 6.10,

$$P\left(\bigcup_{n=1}^{\infty} A_n\right) \leqslant \sum_{n=1}^{\infty} P(A_n) = 0$$

if $P(A_n) = 0$ for $n = 1, 2, \ldots$ .

**6.12** We have $P(\Omega \setminus A_n) = 0$ for $n = 1, 2, \ldots$ , since $P(A_n) = 1$. Then, by de Morgan's law and Problem 6.11,

$$1 - P\left(\bigcap_{n=1}^{\infty} A_n\right) = P\left(\Omega \setminus \bigcap_{n=1}^{\infty} A_n\right) = P\left(\bigcup_{n=1}^{\infty} (\Omega \setminus A_n)\right) = 0,$$

which proves the assertion.

**6.13** From Problem 4.24, we have $|P(A_n) - P(A)| \leqslant P(A_n \triangle A) \to 0$, which implies that $P(A_n) \to P(A)$ as $n \to \infty$.

**6.14** Since

$$\limsup_{n\to\infty} A_n \subset \bigcup_{n=k}^{\infty} A_n$$

for any $k = 1, 2, \ldots$ , it follows that

$$\begin{aligned} P(\limsup_{n\to\infty} A_n) &\leqslant P\left(\bigcup_{n=k}^{\infty} A_n\right) && \text{by Problem 4.12} \\ &\leqslant \sum_{n=k}^{\infty} P(A_n) && \text{by Problem 6.10} \\ &\to 0 \quad \text{as } k \to \infty && \text{if } \sum_{n=1}^{\infty} P(A_n) < \infty. \end{aligned}$$

This completes the proof.

**6.15** Consider

$$B_n = \bigcap_{k=n}^{\infty} A_k, \qquad C_n = \bigcup_{k=n}^{\infty} A_k.$$

Then

$$\begin{aligned} P\left(\liminf_{n\to\infty} A_n\right) &= P\left(\bigcup_{n=1}^{\infty} B_n\right) && \text{by Definition 5.4} \\ &= \lim_{n\to\infty} P(B_n) && \text{since } B_1 \subset B_2 \subset \cdots \\ &= \liminf_{n\to\infty} P(B_n) && \text{since } P(B_1) \leqslant P(B_2) \leqslant \cdots \\ &\leqslant \liminf_{n\to\infty} P(A_n) && \text{since } B_n \subset A_n. \end{aligned}$$

In a similar way, using $C_n$ in place of $B_n$, we can show that

$$\limsup_{n\to\infty} P(A_n) \leqslant P(\limsup_{n\to\infty} A_n).$$

Obviously, $\liminf_{n\to\infty} P(A_n) \leqslant \limsup_{n\to\infty} P(A_n)$, so the proof is completed.

**6.16** $m([\frac{1}{2}, \frac{2}{3})) = \frac{2}{3} - \frac{1}{2} = \frac{1}{6}$,
$m((\frac{1}{5}, \frac{1}{3}]) = \frac{1}{3} - \frac{1}{5} = \frac{2}{15}$,
$m(\{1\}) = m([1, 1]) = 1 - 1 = 0$,
$m(\{\frac{1}{n} : n \in \mathbb{N}\}) = \sum_{n=1}^{\infty} m(\{\frac{1}{n}\}) = 0$,
$m(\mathbb{N}) = \sum_{n=1}^{\infty} m(\{n\}) = 0$.

**6.18** For any fixed $n \in \mathbb{N}$, solving the inequality $|x - n| < \frac{1}{2^n}$, we have $-\frac{1}{2^n} < x - n < \frac{1}{2^n}$, which gives

$$n - \frac{1}{2^n} < x < n + \frac{1}{2^n}, \quad x \in \left(n - \frac{1}{2^n}, n + \frac{1}{2^n}\right).$$

The intervals are pairwise disjoint and have lengths $\frac{1}{2^{n-1}}$, so the measure of the whole set is $\sum_{n=1}^{\infty} \frac{1}{2^{n-1}} = 2$.

**6.20** Since the Cantor set $C$ is of the form $C = \bigcap_{n=1}^{\infty} C_n$, where $C_{n+1} \subset C_n$ (see the solution to Problem 5.23), $m(C) = \lim_{n\to\infty} m(C_n)$ by Problem 6.9. But $m(C_n) = \left(\frac{2}{3}\right)^n$, so $m(C) = \lim_{n\to\infty} \left(\frac{2}{3}\right)^n = 0$.

**6.21** The function is zero outside $[1,3]$, so the integral reduces to this interval: $1 = \int_{\mathbb{R}} c\mathbb{I}_{[1,3]}\,\mathrm{d}m = c\int_1^3 1\,\mathrm{d}x = 2c$; that is, $c = \frac{1}{2}$.

**6.22** The function vanishes for $x$ outside $[a,b]$, so $1 = \int_{\mathbb{R}} c\mathbb{I}_{[a,b]}\,\mathrm{d}m = c\int_a^b \mathrm{d}x = c(b-a)$ and, therefore, $c = \frac{1}{b-a}$.

**6.23** Since

$$\int_{\mathbb{R}} f\,\mathrm{d}m = \int_0^1 cx^n\,\mathrm{d}x = \frac{c}{n+1},$$

we need $c = n+1$ for the integral to be 1.

**6.24** We have

$$\begin{aligned}\int_{\mathbb{R}} f\,\mathrm{d}m &= \int_0^{\infty} ce^{-\lambda x}\,\mathrm{d}x = \lim_{n\to\infty}\int_0^n ce^{-\lambda x}\,\mathrm{d}x \\ &= \lim_{n\to\infty} c\left(-\frac{1}{\lambda}\right)[e^{-\lambda n} - 1] = \frac{c}{\lambda},\end{aligned}$$

so we put $c = \lambda$.

**6.28** Clearly, because $f$ is a density,

$$P(\mathbb{R}) = \int_{\mathbb{R}} f\,\mathrm{d}m = 1.$$

Let us take a sequence of pairwise disjoint Borel sets $A_1, A_2, \ldots$ and put

$$B_n = A_1 \cup \cdots \cup A_n, \qquad B = \bigcup_{i=1}^{\infty} A_i.$$

Since $f\mathbb{I}_{B_n} \nearrow f\mathbb{I}_B$, by (c) of Definition 6.3

$$P(B_n) = \int_{B_n} f\,\mathrm{d}m = \int f\mathbb{I}_{B_n}\,\mathrm{d}m \nearrow \int f\mathbb{I}_B\,\mathrm{d}m = \int_B f\,\mathrm{d}m = P(B),$$

which proves that $P$ is countably additive. It follows that $P$ is a probability measure.

**6.29** By Definition 6.5,

$$P([0,2)) = \int_{[0,2)} \frac{1}{2}\mathbb{I}_{[1,3]}\,\mathrm{d}m = \frac{1}{2}\int_0^1 0\,\mathrm{d}x + \frac{1}{2}\int_1^2 1\,\mathrm{d}x = \frac{1}{2}.$$

**6.30** By Definition 6.5,

$$P([\tfrac{1}{2},2]) = \int_{\frac{1}{2}}^{2}(x-[x])\,\mathrm{d}x = \int_{\frac{1}{2}}^{1} x\,\mathrm{d}x + \int_{1}^{2}(x-1)\,\mathrm{d}x$$
$$= \frac{1}{2}x^2\Big|_{\frac{1}{2}}^{1} + \frac{1}{2}(x-1)^2\Big|_{1}^{2} = \frac{7}{8}.$$

**6.31** $P(\mathbb{N}) = \sum_{k=1}^{\infty} P(\{k\})$ by Definition 6.1 and $P(\{k\}) = \int_k^k f(x)\,\mathrm{d}x = 0$, so $P(\mathbb{N}) = 0$.

**6.34** To verify condition (a) of Definition 6.1, note that $\delta_\omega(\Omega) = 1$ because $\omega \in \Omega$. To verify (b), take a sequence of pairwise disjoint sets $A_i$. Then, either (i) $\omega \notin \bigcup_i A_i$ or (ii) $\omega \in \bigcup_i A_i$.

In case (i), $\delta_\omega(\bigcup_i A_i) = 0$ and $\delta_\omega(A_i) = 0$ for each $i$, so $\sum_i \delta_\omega(A_i) = 0$, which gives (b) of Definition 6.1.

In case (ii), $\omega$ belongs to exactly one of the sets $A_1, A_2, \ldots$ . Suppose that it belongs to $A_k$. Now, $\delta_\omega(\bigcup_i A_i) = 1$, $\delta_\omega(A_k) = 1$, and $\delta_\omega(A_i) = 0$ if $i \neq k$, so $\sum_i \delta_\omega(A_i) = \delta_\omega(A_k) = 1$, which also gives (b).

The largest set of measure 0 is $\Omega \setminus \{\omega\}$. The smallest set of measure 1 is $\{\omega\}$.

**6.35** $P(\Omega) = \frac{1}{2}\delta_{\omega_1}(\Omega) + \frac{1}{2}\delta_{\omega_2}(\Omega) = \frac{1}{2}\times 1 + \frac{1}{2}\times 1 = 1$, so (a) of Definition 6.1 holds.

If $A_1, A_2, \ldots$ is a sequence of disjoint sets, then

$$\begin{aligned}
P\left(\bigcup_{i=1}^{\infty} A_i\right) &= \frac{1}{2}\delta_{\omega_1}\left(\bigcup_{i=1}^{\infty} A_i\right) + \frac{1}{2}\delta_{\omega_2}\left(\bigcup_{i=1}^{\infty} A_i\right) && \text{by the definition of } P \\
&= \frac{1}{2}\sum_{i=1}^{\infty}\delta_{\omega_1}(A_i) + \frac{1}{2}\sum_{i=1}^{\infty}\delta_{\omega_2}(A_i) && \text{since } P_1 \text{ and } P_2 \text{ are measures} \\
&= \sum_{i=1}^{\infty}\left(\frac{1}{2}\delta_{\omega_1}(A_i) + \frac{1}{2}\delta_{\omega_2}(A_i)\right) \\
&= \sum_{i=1}^{\infty} P(A_i),
\end{aligned}$$

so (b) of Definition 6.1 holds too.

The largest set of measure zero is $\Omega \setminus \{\omega_1, \omega_2\}$. Every subset of this set is also of measure zero.

**6.36** First, the usual conditions $\alpha_i \in [0,1]$, and $\sum \alpha_i = 1$ for $P(\Omega)$ to be 1. Under these conditions, $P$ is countably additive:

$$
\begin{aligned}
P\left(\bigcup_{i=1}^{\infty} A_i\right) &= \sum_{k=1}^{\infty} \alpha_k \delta_{\omega_k}\left(\bigcup_{i=1}^{\infty} A_i\right) && \text{by the definition of } P \\
&= \sum_{k=1}^{\infty} \alpha_k \left(\sum_{i=1}^{\infty} \delta_{\omega_k}(A_i)\right) && \text{since } \delta_{\omega_k} \text{ are measures} \\
&= \sum_{i=1}^{\infty} \left(\sum_{k=1}^{\infty} \alpha_k \delta_{\omega_k}(A_i)\right) && \text{since all terms are non-negative} \\
&= \sum_{i=1}^{\infty} P(A_i).
\end{aligned}
$$

Since $\alpha_i \geqslant 0$ and $\sum_{i=1}^{\infty} \alpha_i < \infty$, the above series are convergent and their terms can be rearranged arbitrarily without affecting the sum.

A set has measure zero if it does not contain any of the $\omega_k$'s or contains only those with the corresponding $\alpha_k = 0$.

**6.37** $P(\Omega) = \alpha_1 P_1(\Omega) + \alpha_2 P_2(\Omega) = \alpha_1 + \alpha_2 = 1$, so $P$ satisfies (a) of Definition 6.1.

If $A_1, A_2, \ldots$ is a sequence of pairwise disjoint events, then

$$
\begin{aligned}
P\left(\bigcup_{i=1}^{\infty} A_i\right) &= \alpha_1 P_1\left(\bigcup_{i=1}^{\infty} A_i\right) + \alpha_2 P_2\left(\bigcup_{i=1}^{\infty} A_i\right) && \text{by the definition of } P \\
&= \alpha_1 \sum_{i=1}^{\infty} P_1(A_i) + \alpha_2 \sum_{i=1}^{\infty} P_2(A_i) && \text{since } P_1 \text{ and } P_2 \text{ are measures} \\
&= \sum_{i=1}^{\infty} (\alpha_1 P_1(A_i) + \alpha_2 P_2(A_i)) \\
&= \sum_{i=1}^{\infty} P(A_i) && \text{again by the definition of } P.
\end{aligned}
$$

Condition (b) of Definition 6.1 is also satisfied for $P$.

**6.38** We have

$$
P(\Omega) = \sum_{k=1}^{\infty} \alpha_k P_k(\Omega) = \sum_{k=1}^{\infty} \alpha_k = 1.
$$

If $A_1, A_2, \ldots$ is a sequence of pairwise disjoint events, then

$$\begin{aligned} P\left(\bigcup_{i=1}^{\infty} A_i\right) &= \sum_{k=1}^{\infty} \alpha_k P_k \left(\bigcup_{i=1}^{\infty} A_i\right) && \text{by the definition of } P \\ &= \sum_{k=1}^{\infty} \alpha_k \sum_{i=1}^{\infty} P_k(A_i) && \text{since } P_k \text{ are measures} \\ &= \sum_{i=1}^{\infty} \sum_{k=1}^{\infty} \alpha_k P_k(A_i) && \text{since all terms are non-negative} \\ &= \sum_{i=1}^{\infty} P(A_i) && \text{again by the definition of } P. \end{aligned}$$

Because $\alpha_i \geqslant 0$ and $\sum_{i=1}^{\infty} \alpha_i < \infty$, the above series are convergent and their terms can be rearranged arbitrarily without affecting the sum.

**6.39** $\delta_0([0,2]) = 1$ and $P_2([0,2]) = \frac{1}{2}$, so $P([0,2]) = \frac{2}{3}$.

**6.42** We shall demonstrate that for some $\varepsilon > 0$, there exists a sequence $x_1, x_2, \ldots \in X$ such that $x_i > \varepsilon$ for all $i = 1, 2, \ldots$ , which is a stronger statement than that to be proved.

Suppose that there is no such $\varepsilon > 0$. It follows that for every $n = 1, 2, \ldots$, the set $A_n = \left\{x \in X : x > \frac{1}{n}\right\}$ is finite. But $X = \bigcup_{n=1}^{\infty} A_n$, which means that $X$ is a countable union of finite sets, contradicting the assumption that $X$ is uncountable. This proves the above statement.

# 7

# Conditional Probability and Independence

## 7.1 Theory and Problems

Our knowledge of whether a certain event has occurred or not may influence the odds about other events. This is illustrated by the following problem. As always, we take $(\Omega, \mathcal{F}, P)$ to be a probability space.

**7.1** A fair die with faces 1, 2, and 3 colored green and faces 4, 5, and 6 colored red is tossed once.

a) How likely is it that the outcome is an even number?

b) If you can see that the die has landed green face up (but cannot see the actual number shown), how likely will it be that the outcome is an even number?

Suppose that that a certain event $B \subset \Omega$ is known to have occurred. What is the probability of another event $A \subset \Omega$ in these circumstances? The solution to Problem 7.1 suggests the following.

▶ **Definition 7.1** Whenever $\Omega$ is finite with uniform probability $P$ and $\emptyset \neq B \subset \Omega$, we define

$$P(A|B) = \frac{\#(A \cap B)}{\#B}.$$

This is called the *conditional probability of $A$ given $B$*.

The above formula for $P(A|B)$ can be obtained by computing the probability of $A \cap B$ in $B$, the latter regarded as a new probability space with uniform probability. However, it will no longer apply in the case of non-uniform probability measure $P$ on $\Omega$. We shall seek a suitable generalization to cover this case and possibly also the case of infinite $\Omega$.

**7.2** Suppose that the die in Problem 7.1 is biased with

$$P(1) = P(3) = P(5) = \frac{1}{9}, \quad P(2) = P(4) = P(6) = \frac{2}{9}.$$

a) How likely is it now that the outcome is an even number?
b) What is the probability that the outcome is an even number given that the die lands green face up?

From the solutions to Problems 7.1 and 7.2, we can extract the rules for computing the probability of an event $A$ if another event $B$ is known to have occurred. This leads to the following general definition of conditional probability.

▶ **Definition 7.2** Let $(\Omega, \mathcal{F}, P)$ be a probability space and $B \in \mathcal{F}$ an event with $P(B) \neq 0$. We call

$$P(A|B) = \frac{P(A \cap B)}{P(B)}$$

the *conditional probability of A given B*.

Conditional probability can also be interpreted in terms of a long sequence of random experiments; see Problem 12.16.

**7.3** Two fair dice are rolled and the outcome is kept secret. You are vitally (financially) interested in the sum shown.
a) What is the probability that the sum will be at least 5?
b) Suppose you have been told that at least one die shows 1. How likely is it now that the sum will be 5 or more?

**7.4** You have heard that an old friend of yours has two children, one of whom is a girl, but you do not know the sex of the other child. How likely is it that it is a boy?

Now some properties of conditional probability. Show that:

**7.5** $P(A|\Omega) = P(A)$.

**7.6** If $B \subset A$ and $P(B) \neq 0$, then $P(A|B) = 1$.

**7.7** If $A \cap B = \emptyset$ and $P(B) \neq 0$, then $P(A|B) = 0$.

**7.8** If $A \cap B = \emptyset$ and $P(C) \neq 0$, then $P(A \cup B|C) = P(A|C) + P(B|C)$.

**7.9** $A \mapsto P(A|B)$ is a probability measure on $\mathcal{F}$.

**7.10** $(B, \{A \cap B : A \in \mathcal{F}\}, P(\cdot\,|B))$ is a probability space.

**7.11** If $P(A \cap B) \neq 0$, then $P(A \cap B \cap C) = P(A)P(B|A)P(C|A \cap B)$.

▶ **7.12** If $P(A_1 \cap \cdots \cap A_{n-1}) \neq 0$, then

$$P(A_1 \cap \cdots \cap A_n) = P(A_1)P(A_2|A_1)P(A_3|A_1 \cap A_2) \cdots \cdots P(A_n|A_1 \cap \cdots \cap A_{n-1}).$$

**7.13** If $P(B \cap C) \neq \emptyset$, then $P(A|B \cap C)P(B|C) = P(A \cap B|C)$.

**7.14** If $P(A) \neq 0$ and $P(B) \neq 0$, then $P(A|B) \geqslant P(A)$ is equivalent to $P(B|A) \geqslant P(B)$.

**7.15** Assuming that $P(B) \neq 0$ and $P(\Omega \setminus B) \neq 0$, if $P(A|B) \geqslant P(A)$, then $P(A|\Omega \setminus B) \leqslant P(A)$.

It is time for problems in which we can apply conditional probability.

**7.16** A student has to sit for an examination consisting of 3 questions selected randomly from a list of 100 questions. To pass, he needs to answer all three questions. What is the probability that the student will pass the examination if he knows the answers to 90 questions on the list?

**7.17** Mrs. Jones has made a steak and kidney pie for her two sons. Eating more than a half of it will give indigestion to anyone. While she is away having tea with a neighbor, the older son helps himself to a piece of the pie. Then, the younger son comes and has a piece of what is left. When Mrs. Jones returns, she finds that more than a half of the pie is gone.

Assuming that the size of each of the two pieces eaten is random and uniformly distributed over what is currently available, what is the probability that neither of Mrs. Jones' sons will get indigestion?

Below we shall derive an important result called the *total probability formula*. We begin with a simple example where the probability space is divided in a natural way into disjoint events.

**7.18** Groping about in the dark, we open one of two drawers in a chest and pick up an item of clothing at random. What is the probability that it is a sock if one drawer contains 6 socks and 6 underpants, and the other contains 2 socks and 4 handkerchiefs?

**7.19** Let $\Omega = H_1 \cup H_2$ and $H_1 \cap H_2 = \emptyset$. Assuming that $P(A|H_1)$, $P(A|H_2)$, $P(H_1) \neq 0$, and $P(H_2) \neq 0$ are known, find an expression for $P(A)$.

▶ **7.20** Let $H_1, \dots, H_n$ be pairwise disjoint events ($H_i \cap H_j = \emptyset$ if $i \neq j$) such that $\Omega = H_1 \cup \dots \cup H_n$ and $P(H_i) \neq 0$ for $i = 1, \dots, n$. Show that for any event $A$,

$$P(A) = P(A|H_1)P(H_1) + \dots + P(A|H_n)P(H_n).$$

This is known as the *total probability formula.* It can be generalized to the case of countably many events. We say that events $H_1, H_2, \dots$ form a *partition* of $\Omega$ if they are pairwise disjoint ($H_i \cap H_j = \emptyset$ if $i \neq j$) and

$$\Omega = \bigcup_{n=1}^{\infty} H_n.$$

▶ **Theorem 7.1** *Let $H_1, H_2, \dots$ be a partition of $\Omega$ such that $P(H_n) \neq 0$ for any $n = 1, 2, \dots$. Then, for any event $A$,*

$$P(A) = \sum_{n=1}^{\infty} P(A|H_n)P(H_n).$$

**7.21** Prove Theorem 7.1.

Apply the total probability formula to solve the following problems:

**7.22** A survey is carried out to find the percentage of men who sing in the bathroom. Because some people may be too embarrassed to admit openly to be bathroom singers, each person questioned is asked to roll a die in secret and answer NO if the number shown is 1 and YES if it is 6, no matter what the true answer is, but to tell the truth (YES or NO) if 2, 3, 4, or 5 comes out. Because the number shown is not revealed, it is impossible to tell from the answer given whether the person is a bathroom singer or not.

Suppose that the probability of answering YES in this survey is found to be $\frac{2}{3}$. What is the probability of being a bathroom singer, then?

**7.23** We have two dice, a red and a white one. The red die is loaded so the probability of getting 6 is $\frac{1}{3}$, the remaining outcomes being equally likely among themselves. The white die is fair. We can pick one die, leaving the other one to the opponent. The dice are rolled and the player with the highest result wins. In case of equal results, the player with the white die wins. Which die does one choose to have a better chance of winning the game?

Suppose that the probability space $\Omega$ is divided into two (or more) disjoint events $H_1$ and $H_2$ for which the conditional probabilities $P(A|H_1)$ and $P(A|H_2)$ of a certain event $A$ are known. The problem is to find $P(H_1|A)$ or $P(H_2|A)$. Let us consider a typical example of this kind, which will lead us to the celebrated *Bayes formula.*

**7.24** An item of clothing is picked at random from one of two drawers in a dark room as described in Problem 7.18. The first drawer contains 6 socks and 6 underpants, and the other contains 2 socks and 4 handkerchiefs. What is the probability that the item comes from the first drawer if it turns out to be a sock?

**7.25** Let $H_1$ and $H_2$ be disjoint events such that $H_1 \cup H_2 = \Omega$. Express $P(H_1|A)$ in terms of $P(H_1)$, $P(H_2)$, $P(A|H_1)$, and $P(A|H_2)$.

**7.26** Generalize Problem 7.25 to the case of $n$ events $H_1, \ldots, H_n$.

The results of Problems 7.25 and 7.26 can be extended to the case of countably many events $H_1, H_2, \ldots$ .

▶ **Theorem 7.2** *Let $H_1, H_2, \ldots$ be a partition of $\Omega$ such that $P(H_n) \neq 0$ for any $n = 1, 2, \ldots$ and let $P(A) \neq 0$. Then,*

$$P(H_1|A) = \frac{P(A|H_1)P(H_1)}{\sum_{n=1}^{\infty} P(A|H_n)P(H_n)}.$$

This is called the *Bayes formula.*

**7.27** Prove the Bayes formula in Theorem 7.2.

**7.28** You have two coins, a fair one with probability of heads $\frac{1}{2}$ and an unfair one with probability of heads $\frac{1}{3}$, but otherwise identical. A coin is selected at random and tossed, falling heads up. How likely is it that it is the fair one?

**7.29** You have three coins in your pocket, two fair ones but the third biased with probability of heads $p$ and tails $1-p$. One coin selected at random drops to the floor, landing heads up. How likely is it that it is one of the fair coins?

**7.30** In the survey in Problem 7.22, the probability of being a bathroom singer is found to be $\frac{3}{4}$. How likely is it that a person who gave a YES answer is indeed a bathroom singer? How likely is it that a person who answered NO is a bathroom singer?

We have learned at the beginning of this chapter that the likelihood of an event $A$ may be affected by the occurrence of another event $B$. However, this is not always the case. Here is an example.

**7.31** A card is drawn at random from a standard deck of 52 playing cards. Let $A$ denote the event that the card is an Ace and $B$ the event that the card is a spade. Show that $P(A) = P(A|B)$.

The condition $P(A) = P(A|B)$ is a key to the notion of *independence*, to which the rest of this chapter will be devoted. Let us study this condition in some detail before stating the formal definition.

**7.32** Suppose that $P(A) \neq 0$ and $P(B) \neq 0$. Show that the following conditions are equivalent:
a) $P(A) = P(A|B)$,
b) $P(B) = P(B|A)$,
c) $P(A)P(B) = P(A \cap B)$.

**7.33** Let $0 \neq P(B) \neq 1$. Show that the following conditions are equivalent:
a) $P(A) = P(A|B)$,
b) $P(A) = P(A|\Omega \setminus B)$,
c) $P(A|B) = P(A|\Omega \setminus B)$.

Among the conditions in Problems 7.32 and 7.33, $P(A \cap B) = P(A)P(B)$ is the only one that requires no additional assumptions such as $P(A) \neq 0$ or $P(B) \neq 0$. Perhaps even more important is the symmetry of this condition: It remains the same if $A$ and $B$ are interchanged.

▶ **Definition 7.3** Events $A$ and $B$ are said to be *independent* if

$$P(A \cap B) = P(A)P(B).$$

**7.34** Show that for any events $A$ and $B$, the following conditions are equivalent:
a) $A$ and $B$ are independent,
b) $\Omega \setminus A$ and $B$ are independent,
c) $\Omega \setminus A$ and $\Omega \setminus B$ are independent.

**7.35** Let $\Omega = \{1, 2, 3, 4\}$ with uniform probability and let $A = \{1, 2\}$. List all $B \subset \Omega$ such that $A$ and $B$ are independent.

**7.36** Let $\Omega = \{1, 2, 3, 4, 5, 6\}$ with uniform probability. Show that if $A, B \subset \Omega$ are independent and $A$ has 4 elements, then $B$ must have 0, 3, or 6 elements.

**7.37** Let $\Omega$ be an $n$-element set with uniform probability and let $A, B \subset \Omega$ be independent. Show that if $A$ has $i$ elements, then $B$ must have

$$j = k\frac{n}{\gcd(i, n)}$$

elements, where $k \in \{0, 1, \ldots, \gcd(i, n)\}$. Here, $\gcd(i, n)$ is the greatest common divisor of $i, n$.

**7.38** Show that if $A = B$ and $A, B$ are independent (that is, $A$ is independent of itself), then $P(A) = 0$ or 1.

**7.39** Let $A$ be an event. Prove that the following conditions are equivalent:
a) $A, B$ are independent for any event $B$,
b) $P(A) = 0$ or 1.

We wish to extend independence to more than two events. First let us investigate the problems below.

**7.40** A fair coin is tossed three times and the following events are considered:
$A =$ toss 1 and toss 2 produce different outcomes,
$B =$ toss 2 and toss 3 produce different outcomes,
$C =$ toss 3 and toss 1 produce different outcomes.
Show that $P(A) = P(A|B) = P(A|C)$, but $P(A) \neq P(A|B \cap C)$.

**7.41** Show that the events $A, B, C$ in Problem 7.40 are pairwise independent; that is, $A, B$ are independent in the sense of Definition 7.3, and so are $B, C$ and $C, A$.

In Problem 7.40, the likelihood of $A$ is not affected by the occurrence of either one of the events $B$ or $C$, but it changes to 0 when $B$ and $C$ occur simultaneously. We shall not consider such three events independent.

For independent events, it is required that the probability of any one of them should not be affected by the occurrence of one or more of the remaining events. Problem 7.41 shows that pairwise independence is insufficient. We need an additional condition.

**7.42** Show that if $A$, $B$, and $C$ are pairwise independent events and $P(B \cap C) \neq 0$, then the following conditions are equivalent:
a) $P(A) = P(A|B \cap C)$,
b) $P(A \cap B \cap C) = P(A)P(B)P(C)$.

▶ **Definition 7.4** Three events $A$, $B$, and $C$ are called *independent* if

(a) they are pairwise independent; that is, $A, B$ are independent, $B, C$ are independent, and $C, A$ are independent in the sense of Definition 7.3;

(b) $P(A \cap B \cap C) = P(A)P(B)P(C)$.

**7.43** Let $A, B, C$ be independent events. Show that
a) $A, B \cap C$ are independent,
b) $A, B \cup C$ are independent,
c) $A, B \setminus C$ are independent.

$*$ **7.44** Let $A$, $B$, and $C$ be independent events. Show that if $D$ is any event in the $\sigma$-field $\mathcal{F}_{\{B,C\}}$ generated by $\{B, C\}$, then $A$ and $D$ are independent.

**7.45** Prove that the following conditions are equivalent:
a) $A, B, C$ are independent,
b) $\Omega \setminus A, B, C$ are independent,
c) $\Omega \setminus A, \Omega \setminus B, C$ are independent,
d) $\Omega \setminus A, \Omega \setminus B, \Omega \setminus C$ are independent.

Finally, we generalize Definition 7.4 to any finite number of events. What follows is an example of definition by induction. Definition 7.3 serves as the base of induction.

▶ **Definition 7.5** We say that events $A_1, \ldots, A_n$ are *independent* if $n = 2$ and $A_1, A_2$ are independent in the sense of Definition 7.3, or $n > 2$ and

(a) for any $k = 1, \ldots, n$, the $n-1$ events $A_1, \ldots, A_{k-1}, A_{k+1}, \ldots, A_n$ are independent;

(b) $P(A_1 \cap \cdots \cap A_n) = P(A_1) \cdots P(A_n)$.

**7.46** Let $A_1, \ldots, A_n, B$ be $n+1$ independent events. Show that $B$ and $C$ are independent for any event $C$ in the $\sigma$-field $\mathcal{F}_{\{A_1,\ldots,A_n\}}$ generated by $\{A_1, \ldots, A_n\}$.

**7.47** Let $A_1, \ldots, A_n, B_1, \ldots, B_m$ be $n+m$ independent events. Show that for any $C$ in $\mathcal{F}_{\{A_1,\ldots,A_n\}}$ and any $D$ in $\mathcal{F}_{\{B_1,\ldots,B_m\}}$, the events $C$ and $D$ are independent.

## 7.2 Hints

**7.1** Because the die is fair, take uniform probability.
a) The solution is rather obvious in this case. All you need to do is to count all favorable outcomes and all possible outcomes.
b) How many favorable outcomes are there, given that the die has landed green face up? How many possible outcomes are there?

**7.2** In the case of a biased die, the probability measure is not uniform, so counting all favorable and possible outcomes will not do any more.
a) You need to add up the known probabilities of smaller disjoint events to find the probability of a larger event.
b) List all possible outcomes and all favorable outcomes such that the die shows a green face.
Consider the three possible outcomes 1, 2, and 3 on your list as a new probability space, on which you will need a suitable probability measure. For the same reason as in a), counting all possible and favorable outcomes on your list will not do for a biased die. But discarding the "red" outcomes 4, 5, and 6 cannot affect the relative odds of the remaining outcomes 1, 2, and 3. What are these relative odds, then?

Now that you know the relative probabilities of the outcomes in the probability space $\{1, 2, 3\}$, you can find the probabilities themselves because they must add up to 1. It remains to use these probabilities to find the likelihood of the set of favorable outcomes on your list.

**7.3** An outcome will be called favorable if the sum shown is at least 5.

a) What probability space $\Omega$ describes all possible outcomes when two dice are rolled? Since the dice are fair, use uniform probability. How many possible outcomes are there? How many are favorable?

b) If it is known that at least one die shows 1, which outcomes in the above $\Omega$ cannot happen? How many are left? How many of these are favorable?

**7.4** A common but false answer is $\frac{1}{2}$ based on the reasoning that it is equally likely that the other child is a boy or a girl. Can you see why this is not so?

Out of the four equally likely possibilities, BB, BG, GB, and GG, how many will be left if you know that at least one child is a girl? Which of the possibilities left mean that the other child is a boy?

**7.5** Just apply the definition of conditional probability. What is $P(\Omega)$? Can you simplify $A \cap \Omega$?

**7.6** Can you simplify $A \cap B$ if $B \subset A$?

**7.7** If $A \cap B = \emptyset$, then what is $P(A \cap B)$?

**7.8** Use the identity $(A \cup B) \cap C = (A \cap C) \cup (B \cap C)$ and the fact that $A \cap C$ and $B \cap C$ are disjoint if $A \cap B = \emptyset$.

**7.9** To show that $P(\Omega|B) = 1$, see Problem 7.6.

To show that the map $A \mapsto P(A|B)$ is countably additive, use the definition of conditional probability and observe that if $A_1, A_2, \ldots$ is a sequence of pairwise disjoint events, then so is $A_1 \cap B, A_2 \cap B, \ldots$ .

**7.10** Refer to Problem 5.9 to show that $\{A \cap B : A \in \mathcal{F}\}$ is a $\sigma$-field. A short cut to verifying countable additivity for $P(\cdot\,|B)$ is through Problem 7.9.

**7.11** You can write the definition of conditional probability as $P(X \cap Y) = P(X)P(Y|X)$. Apply this formula twice, substituting suitable events for $X$ and $Y$.

**7.12** Refer to Problem 7.11 for $n = 3$. For larger $n$, use the definition of conditional probability repeatedly or prove the formula by induction.

**7.13** Observe that $P(B \cap C) \neq 0$ implies that $P(C) \neq 0$, and apply the definition of conditional probability.

**7.14** Just use the definition of conditional probability.

**7.15** If $P(A|B)$, $P(A|\Omega \setminus B)$, and $P(B)$ are known, can you compute $P(A)$? To this end, you need to express $A$ using $A \cap B$ and $A \cap (\Omega \setminus B)$. Having written $P(A)$ in terms of $P(A|B)$, $P(A|\Omega \setminus B)$, and $P(B)$, try to obtain the desired inequality.

**7.16** What is the probability that the student can answer the first exam question? What is the conditional probability that he can answer the second question given that he can answer the first one? What is the conditional probability that he can answer the third question given that he can answer the first two? Now, use the formula in Problem 7.11.

**7.17** All possible outcomes can be described as pairs of numbers, namely the portions of the pie eaten by the two sons. It helps to draw the set $\Omega$ of all possible outcomes as a subset of the plane. Observe that the older son can take as much as he wants from the whole pie, but the younger son's choice is restricted to what is left by his elder brother. Also, draw the two events: $B$, that Mrs. Jones will find more than a half of the pie gone, and $A$, that neither of her sons will get indigestion (that is, eat more than a half of the whole pie).

Now, you need to introduce a probability measure on $\Omega$ consistent with the assumption that the size of each of the two pieces eaten is random and uniformly distributed over what is currently available to either of the two sons. This can be achieved by defining a suitable density over $\Omega$.

Once this is done, you can compute the probability of any event in $\Omega$ by integration. This will enable you to find the conditional probability $P(A|B)$ from definition.

**7.18** Consider the following events:

$$\begin{aligned} S &= \text{the item picked is a sock}, \\ D_1 &= \text{the item comes from the first drawer}, \\ D_2 &= \text{the item comes from the second drawer}. \end{aligned}$$

What is the probability $P(D_1)$ that the item comes from the first drawer? What is the conditional probability $P(S|D_1)$ that the item is a sock given that it comes from the first drawer? What is $P(S \cap D_1)$, then? Now, answer similar questions for the other drawer. To find $P(S)$, observe that $S = (S \cap D_1) \cup (S \cap D_2)$.

There is no need to specify the probability space $\Omega$ explicitly.

**7.19** See the solution to Problem 7.18.

**7.20** The solution to Problem 7.19 extends easily to $n$ sets $H_1, \ldots, H_n$.

**7.21** The solution to Problem 7.20 extends easily to countably many events $H_1, H_2, \ldots$ .

**7.22** What is the conditional probability that the person questioned answers YES given that the die shows 1? What is the conditional probability of a YES answer if the outcome is 6? What if the outcome is 2, 3, 4, or 5?

Now, use the total probability formula to express the probability of a YES answer in terms of the above conditional probabilities, one of which is related to the likelihood of being a bathroom singer.

**7.23** At least two different partitions of the probability space come to mind:

$$W_1, \ldots, W_6 \quad \text{or} \quad R_1, \ldots, R_6,$$

where $W_i$ is the event that the white die shows $i$ and $R_i$ that the red die shows $i$. Either can be used in the total probability formula to compute the likelihood of winning the game. Which partition do you think will be easier to use?

**7.24** Consider the same events

$$\begin{aligned} S &= \text{the item picked is a sock}, \\ D_1 &= \text{the item comes from the first drawer}, \\ D_2 &= \text{the item comes from the second drawer} \end{aligned}$$

as in Problem 7.18, where $P(S)$ was found. What else do you need to know to compute the probability $P(D_1|S)$?

**7.25** See Problem 7.24.

**7.26** As compared to Problem 7.25, the only difference is that the total probability formula for $n$ events $H_1, \ldots, H_n$ should be used here.

**7.27** See Problems 7.25 and 7.26 and use Theorem 7.1

**7.28** Use Bayes' formula to express $P(F|H)$.

Consider the following events:

$$\begin{aligned} F &= \text{the selected coin is fair}, \\ U &= \text{the selected coin is unfair}, \\ H &= \text{the coin lands heads up}. \end{aligned}$$

What is the probability $P(F)$ of selecting the fair coin? What is the conditional probability $P(H|F)$ of heads given that the selected coin is fair? Now, answer the same questions for the unfair coin.

**7.29** See Problem 7.28, but some probabilities will change.

**7.30** Imagine you are a bathroom singer yourself. If you roll a die and answer according to the rules of the survey described in Problem 7.22, how likely is it that your answer will be YES? This gives $P(Y|S)$.

Now imagine you are not a bathroom singer. What is the probability that you will answer YES? This gives $P(Y|\Omega \setminus S)$.

It remains to apply the Bayes formula to find $P(S|Y)$.

**7.31** How many possible outcomes are there? How many of these are in $A$? This will give you $P(A)$. To find $P(A|B)$, you need to count the elements of $A \cap B$ and $B$.

**7.32** Apply the definition of conditional probability.

**7.33** Write $P(A|\Omega \setminus B)$ and $P(A|B)$ in terms of $P(A)$, $P(B)$, and $P(A \cap B)$.

**7.34** It suffices to prove that a) $\Leftrightarrow$ b). All you need is Definition 7.3.

**7.35** Here, $A$ and $\Omega \setminus A$ have the same number of elements. In this case, $A$ and $B$ will be independent if and only if $A \cap B$ and $(\Omega \setminus A) \cap B$ have the same number of elements. List all such sets $B$.

**7.36** Let $j$ be the number of elements in $B$. You want to prove that $j$ is divisible by 3. Observe that $A$ and $B$ are independent when $4j = 6m$, where $m$ is the number of elements in $A \cap B$.

**7.37** You want to prove that $j$ is divisible by $\frac{n}{\gcd(i,n)}$. Observe that $A$ and $B$ are independent when $ij = mn$, where $m$ is the number of elements of $A \cap B$.

**7.38** Show that P(A) satisfies the equation $x(x-1) = 0$.

**7.39** If $P(A)$ is 0, then what is $P(A \cap B)$? If $P(A)$ is 1, then what is $P(A \cap B)$?

**7.40** How many possible outcomes are there if a coin is tossed three times? How many of these are in $A$, in $B$, and in $C$? How many outcomes are in $A \cap B$, $B \cap C$, and $C \cap A$? How many are in $A \cap B \cap C$?

**7.41** Use the probabilities computed in Problem 7.40.

**7.42** Condition a) means that $A$ and $B \cap C$ are independent events. You will also need the independence of $B$ and $C$.

**7.43** Condition a) follows from Problem 7.42. In b), use Problem 4.18. In c), you will need Problem 4.10.

**7.44** Show that the $\sigma$-field $\mathcal{F}_{\{B,C\}}$ generated by $\{B, C\}$ is, in fact, a finite field. What are the atoms of $\mathcal{F}_{\{B,C\}}$? Is each of them independent of $A$?

**7.45** It suffices to show that a) $\Leftrightarrow$ b). The other equivalences follow from this.

**7.46** Consider the family of all events $C$ such that $B$ and $C$ are independent. Is it a $\sigma$-field? Does it contain $\mathcal{F}_{\{A_1,\ldots,A_n\}}$?

**7.47** Consider the family of all events $D$ such that $C$ and $D$ are independent for any $C$ in $\mathcal{F}_{\{A_1,\ldots,A_n\}}$. Is this family a $\sigma$-field? Does it contain $\mathcal{F}_{\{B_1,\ldots,B_m\}}$?

## 7.3 Solutions

**7.1** a) Take

$$\Omega = \{1, 2, 3, 4, 5, 6\}$$

with uniform probability $P$ on $\Omega$. Then,

$$A = \{2, 4, 6\}$$

is the set of favorable outcomes, i.e., even numbers. Hence,

$$P(A) = \frac{\#A}{\#\Omega} = \frac{3}{6} = \frac{1}{2}.$$

b) Given that the die has landed green face up, the set of possible outcomes is

$$B = \{1, 2, 3\}.$$

In this case, the set of favorable outcomes is just

$$A \cap B = \{2\}.$$

Because nothing else is known to distinguish between the three outcomes in $B$, they should be equally likely. This means that we need to take the uniform probability measure $\tilde{P}$ on $B$. Thus, the desired probability that the outcome is an even number given that the die lands green face up will be

$$\tilde{P}(A \cap B) = \frac{\#(A \cap B)}{\#B} = \frac{1}{3}.$$

**7.2** a) The set $\Omega = \{1, 2, 3, 4, 5, 6\}$ of all possible outcomes and the set $A = \{2, 4, 6\}$ of favorable outcomes are the same as in the solution to part a) of Problem 7.1. But now

$$P(A) = P(\{2\}) + P(\{4\}) + P(\{6\}) = \frac{2}{9} + \frac{2}{9} + \frac{2}{9} = \frac{2}{3}.$$

b) Given that the die shows a green face, the set of possible outcomes is $B = \{1, 2, 3\}$ and the set of favorable outcomes is $A \cap B = \{2\}$, as in part b) of Problem 7.1.

We need a suitable probability measure $\tilde{P}$ on $B$. Because discarding the "red" outcomes 4, 5, and 6 will not affect the relative odds of the remaining outcomes 1, 2, and 3, we have

$$\tilde{P}(\{1\}) : \tilde{P}(\{2\}) : \tilde{P}(\{3\}) = P(\{1\}) : P(\{2\}) : P(\{3\}) = \frac{1}{9} : \frac{2}{9} : \frac{1}{9}.$$

But $\tilde{P}(B) = 1$, so

$$\tilde{P}(\{1\}) + \tilde{P}(\{2\}) + \tilde{P}(\{3\}) = 1.$$

It follows that $\tilde{P}(\{i\}) = \frac{P(\{i\})}{P(B)}$ for $i = 1, 2, 3$, so

$$\tilde{P}(C) = \frac{P(C)}{P(B)}$$

for any $C \subset B$.

Therefore, the probability that the outcome is an even number given that the die lands green face up is

$$\tilde{P}(A \cap B) = \frac{P(A \cap B)}{P(B)} = \frac{\frac{2}{9}}{\frac{1}{9} + \frac{2}{9} + \frac{1}{9}} = \frac{1}{2}.$$

**7.3** a) When two fair dice are rolled, the set of all possible outcomes is

$$\Omega = \{(i, j) : i, j = 1, \ldots, 6\}$$

and is equipped with uniform probability $P$. We want to compute the probability of the event

$$A = \{(i, j) \in \Omega : i + j \geqslant 5\}.$$

Since $\#\Omega = 36$ and $\#A = 30$, we find that

$$P(A) = \frac{\#A}{\#\Omega} = \frac{30}{36} = \frac{5}{6}.$$

b) We know that at least one die shows 1, which means that the event

$$B = \{(1, j) : j = 1, \ldots, 6\} \cup \{(i, 1) : i = 1, \ldots, 6\}$$

has occurred. Then,

$$A \cap B = \{(1,4),(1,5),(1,6),(4,1),(5,1),(6,1)\}.$$

From Definition 7.1,

$$P(A|B) = \frac{\#(A \cap B)}{\#B} = \frac{6}{11}.$$

**7.4** As the probability space take

$$\Omega = \{\text{BB}, \text{BG}, \text{GB}, \text{GG}\}$$

with uniform probability $P$. Consider the events

$$X = \{\text{BB}, \text{BG}, \text{GB}\}, \quad Y = \{\text{BG}, \text{GB}, \text{GG}\}.$$

Clearly, $X$ occurs if at least one child is a boy and $Y$ occurs if at least one child is a girl.

You want to find the probability that your friend has a boy given that at least one child is a girl, i.e., the conditional probability

$$P(X|Y) = \frac{P(X \cap Y)}{P(Y)} = \frac{\#\{\text{BG}, \text{GB}\}}{\#\{\text{BG}, \text{GB}, \text{GG}\}} = \frac{2}{3}.$$

**7.5** Since $A \cap \Omega = A$ and $P(\Omega) = 1$,

$$P(A|\Omega) = \frac{P(A \cap \Omega)}{P(\Omega)} = P(A).$$

**7.6** If $B \subset A$, then $A \cap B = B$. Hence,

$$P(A|B) = \frac{P(A \cap B)}{P(B)} = \frac{P(B)}{P(B)} = 1.$$

**7.7** If $A \cap B = \emptyset$, then $P(A \cap B) = 0$, so

$$P(A|B) = \frac{P(A \cap B)}{P(B)} = 0.$$

**7.8** If $A \cap B = \emptyset$, then $(A \cap C) \cap (B \cap C) = \emptyset$, so

$$\begin{aligned} P((A \cup B) \cap C) &= P((A \cap C) \cup (B \cap C)) \\ &= P(A \cap C) + P(B \cap C). \end{aligned}$$

Hence,

$$\begin{aligned} P(A \cup B|C) &= \frac{P((A \cup B) \cap C)}{P(C)} \\ &= \frac{P(A \cap C)}{P(C)} + \frac{P(B \cap C)}{P(C)} = P(A|C) + P(B|C). \end{aligned}$$

**7.9** Since $B \subset \Omega$, we have $P(\Omega|B) = 1$ from Problem 7.6.

Now, let $A_1, A_2, \ldots \in \mathcal{F}$ be a sequence of pairwise disjoint events. Then, $A_1 \cap B, A_2 \cap B, \ldots \in \mathcal{F}$ is also a sequence of pairwise disjoint events, so

$$\begin{aligned} P\left(\bigcup_{n=1}^{\infty} A_n \middle| B\right) &= \frac{1}{P(B)} P\left(B \cap \bigcup_{n=1}^{\infty} A_n\right) \\ &= \frac{1}{P(B)} P\left(\bigcup_{n=1}^{\infty} (A_n \cap B)\right) \\ &= \frac{1}{P(B)} \sum_{n=1}^{\infty} P(A_n \cap B) \\ &= \sum_{n=1}^{\infty} \frac{P(A_n \cap B)}{P(B)} \\ &= \sum_{n=1}^{\infty} P(A_n|B). \end{aligned}$$

**7.10** From Problem 5.9, we know that $\{A \cap B : A \in \mathcal{F}\}$ is a $\sigma$-field.

By Problem 7.9, $P(\,\cdot\,|B)$ is countably additive on $\mathcal{F}$ (condition (b) of Definition 6.1), so it must also be countably additive on the smaller $\sigma$-field $\{A \cap B : A \in \mathcal{F}\}$. But $P(B|B) = 1$ by Problem 7.6, so $P(\,\cdot\,|B)$ is a probability measure on $\{A \cap B : A \in \mathcal{F}\}$.

**7.11** From Definition 7.2, we have

$$P(X \cap Y) = P(X)P(Y|X)$$

for any events $X$ and $Y$ with $P(X) \neq 0$. We shall use this formula twice. First, substituting $A \cap B$ for $X$ and $C$ for $Y$, we obtain

$$P(A \cap B \cap C) = P(A \cap B)P(C|A \cap B).$$

Then, taking $A$ in place of $X$ and $B$ in place of $Y$, we get

$$P(A \cap B) = P(A)P(B|A).$$

Substituting this for $P(A \cap B)$ above, we obtain the desired formula.

**7.12** We shall prove the formula by induction on $n$. The definition of conditional probability written as $P(A_1 \cap A_2) = P(A_1)P(A_2|A_1)$ serves as the induction base.

Now, suppose that the formula to be proved holds for some fixed $n$. This will be our induction hypothesis. We shall demonstrate that the formula holds for $n + 1$.

Let us take $n+1$ events $A_1, A_2, \ldots, A_n, A_{n+1}$ such that $P(A_1 \cap A_2 \cap \cdots \cap A_n) \neq 0$. Considering $A_1 \cap A_2$ as a single event, we can apply the induction hypothesis to the $n$ events $A_1 \cap A_2, A_3, \ldots, A_n$, obtaining

$$P((A_1 \cap A_2) \cap A_3 \cap \cdots \cap A_n) = P(A_1 \cap A_2) P(A_3 | A_1 \cap A_2) \cdots \\ \cdots P(A_n | A_1 \cap A_2 \cap A_3 \cap \cdots \cap A_{n-1}).$$

But $P(A_1 \cap A_2) = P(A_1) P(A_2 | A_1)$ by the definition of conditional probability. Substituting this into the above expression, we get

$$P(A_1 \cap A_2 \cap \cdots \cap A_n) = P(A_1) P(A_2 | A_1) P(A_3 | A_1 \cap A_2) \cdots \\ \cdots P(A_n | A_1 \cap A_2 \cap \cdots \cap A_{n-1}),$$

completing the proof.

**7.13** Since $P(B \cap C) \neq 0$, it follows that $P(C) \neq 0$ because $B \cap C \subset C$. By the definition of conditional probability,

$$\begin{aligned} P(A|B \cap C) P(B|C) &= \frac{P(A \cap B \cap C)}{P(B \cap C)} \frac{P(B \cap C)}{P(C)} \\ &= \frac{P(A \cap B \cap C)}{P(C)} \\ &= P(A \cap B | C). \end{aligned}$$

**7.14** Since

$$P(A|B) = \frac{P(A \cap B)}{P(B)}, \quad P(B|A) = \frac{P(A \cap B)}{P(A)}$$

by definition, it follows that $P(A|B) \geqslant P(A)$ if and only if $P(A \cap B) \geqslant P(A)P(B)$, which, in turn, is equivalent to $P(B|A) \geqslant P(B)$.

**7.15** Since $A = (A \cap B) \cup (A \cap (\Omega \setminus B))$ and the sets $A \cap B$ and $A \cap (\Omega \setminus B)$ are disjoint,

$$\begin{aligned} P(A) &= P(A \cap B) + P(A \cap (\Omega \setminus B)) \\ &= P(A|B)P(B) + P(A|\Omega \setminus B) P(\Omega \setminus B). \end{aligned}$$

If $P(A|B) \geqslant P(A)$, it follows that

$$\begin{aligned} P(A|\Omega \setminus B) P(\Omega \setminus B) &= P(A) - P(A|B)P(B) \\ &\leqslant P(A) - P(A)P(B) \\ &= P(A)(1 - P(B)) \\ &= P(A) P(\Omega \setminus B). \end{aligned}$$

Dividing both sides by $P(\Omega \setminus B) \neq 0$, we find that $P(A|\Omega \setminus B) \leqslant P(A)$, as required.

**7.16** Let the questions be numbered 1, 2, and 3. By $A_i$ we denote the event that question $i$ is among those the student can answer. We need to compute $P(A_1 \cap A_2 \cap A_3)$.

By Problem 7.11,

$$P(A_1 \cap A_2 \cap A_3) = P(A_1)P(A_2|A_1)P(A_3|A_1 \cap A_2).$$

Because the student can answer 90 out of 100 questions,

$$P(A_1) = \frac{90}{100}.$$

If question 1 is among those the student can answer, question 2 will be one of the remaining 99 questions, out of which the student can answer 89. Hence,

$$P(A_2|A_1) = \frac{89}{99}.$$

If the student can answer questions 1 and 2, then question 3 will be one of the remaining 98 questions, 88 of which can be answered by the student. Thus,

$$P(A_3|A_1 \cap A_2) = \frac{88}{98}.$$

It follows that

$$P(A_1 \cap A_2 \cap A_3) = \frac{90}{100} \times \frac{89}{99} \times \frac{88}{98} \approx 0.73\,.$$

**7.17** Let the whole pie be represented by the interval $[0,1]$ and let $x \in [0,1]$ be the portion eaten by the older son. Then, $[0, 1-x]$ will be available to the younger son, who takes a portion of size $y \in [0, 1-x]$. The set of all possible outcomes $(x, y)$ is

$$\Omega = \{(x,y) : x, y \geqslant 0, x + y \leqslant 1\}\,.$$

Between them, the two sons will have eaten $x + y$, so

$$B = \left\{(x,y) \in \Omega : x + y > \frac{1}{2}\right\}$$

is the event that Mrs. Jones will find more than a half of her pie gone. Furthermore,

$$A = \left\{(x,y) \in \Omega : x, y < \frac{1}{2}\right\}$$

is the event that neither of Mrs. Jones' sons will get indigestion. Of course,

$$A \cap B = \left\{(x,y) \in \Omega : x, y < \frac{1}{2}, x + y > \frac{1}{2}\right\}.$$

The events $A$, $B$, and $A \cap B$ are shown in Figure 7.1.

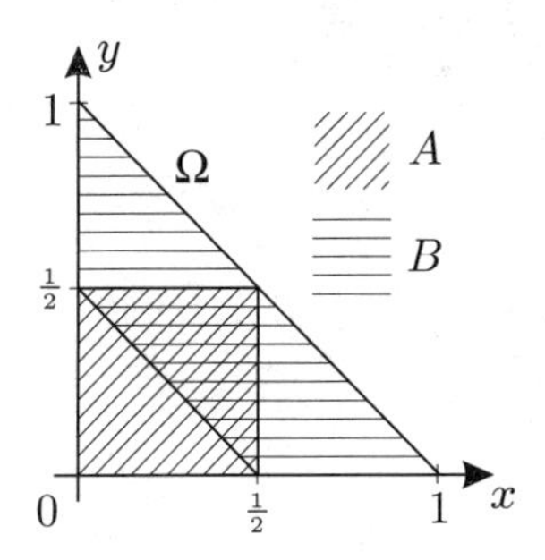


FIGURE 7.1. Events $A$, $B$, and $A \cap B$

If $x$ is uniformly distributed over $[0,1]$ and $y$ is uniformly distributed over $[0,1-x]$, the probability measure $P$ over $\Omega$ with density

$$\rho(x,y) = \frac{1}{1-x}$$

will describe the joint distribution of outcomes $(x,y)$. The density $\rho(x,y)$ is shown in Figure 7.2.

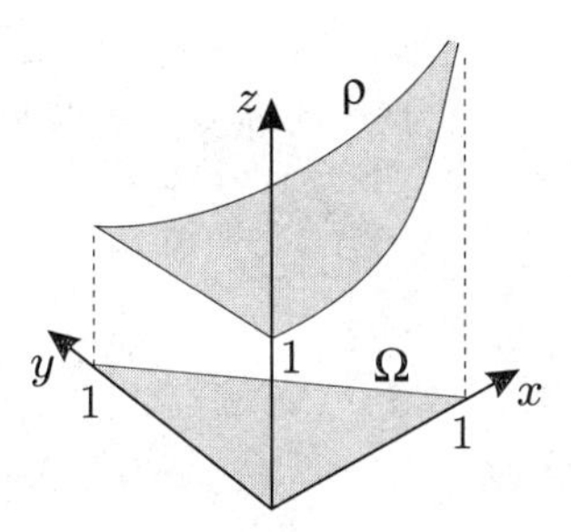


FIGURE 7.2. Density $\rho(x,y) = \frac{1}{x-1}$ over $\Omega$

Let us find $P(B)$:

$$\begin{aligned}
P(B) &= \int_{(x,y)\in B} \rho(x,y)\,\mathrm{d}x\,\mathrm{d}y \\
&= \int_0^{\frac{1}{2}} \left( \int_{\frac{1}{2}-x}^{1-x} \frac{1}{1-x}\,\mathrm{d}y \right) \mathrm{d}x + \int_{\frac{1}{2}}^{1} \left( \int_0^{1-x} \frac{1}{1-x}\,\mathrm{d}y \right) \mathrm{d}x \\
&= \frac{1}{2} \int_0^{\frac{1}{2}} \frac{1}{1-x}\,\mathrm{d}x + \int_{\frac{1}{2}}^{1} \mathrm{d}x \\
&= -\frac{1}{2} \ln(1-x)\Big|_0^{\frac{1}{2}} + \frac{1}{2} = -\frac{1}{2} \ln \frac{1}{2} + \frac{1}{2} \\
&= \ln \sqrt{2e}.
\end{aligned}$$

Next, we find $P(A \cap B)$:

$$\begin{aligned}
P(A \cap B) &= \int_{(x,y)\in A\cap B} \rho(x,y)\,\mathrm{d}x\,\mathrm{d}y \\
&= \int_0^{\frac{1}{2}} \left( \int_{\frac{1}{2}-x}^{\frac{1}{2}} \frac{1}{1-x}\,\mathrm{d}y \right) \mathrm{d}x = \int_0^{\frac{1}{2}} \frac{x}{1-x}\,\mathrm{d}x \\
&= -\ln(1-x)\Big|_0^{\frac{1}{2}} - \frac{1}{2} = -\ln \frac{1}{2} - \frac{1}{2} \\
&= \ln \sqrt{\frac{4}{e}}.
\end{aligned}$$

It follows that

$$P(A|B) = \frac{P(A \cap B)}{P(B)} = \frac{\ln \sqrt{\frac{4}{e}}}{\ln \sqrt{2e}} \approx 0.23\ .$$

**7.18** Consider the events $S$, $D_1$, and $D_2$ introduced in the hint. Assuming that it is equally likely to open either drawer and equally likely to pick any item among those in the open drawer, we have

$$P(D_1) = P(D_2) = \tfrac{1}{2},$$
$$P(S|D_1) = \tfrac{6}{12} = \tfrac{1}{2}, \quad P(S|D_2) = \tfrac{2}{6} = \tfrac{1}{3}.$$

But the events $D_1, D_2$ are disjoint (we open only one drawer) and their union is the whole probability space, $D_1 \cup D_2 = \Omega$ (we do open a drawer). Then, $S \cap D_1$ and $S \cap D_2$ are also disjoint and

$$\begin{aligned}
P(S) &= P(S \cap (D_1 \cup D_2)) && \text{since } D_1 \cup D_2 = \Omega \\
&= P((S \cap D_1) \cup (S \cap D_2)) \\
&= P(S \cap D_1) + P(S \cap D_2) && \text{by additivity} \\
&= P(S|D_1)P(D_1) + P(S|D_2)P(D_2) && \text{by Definition 7.2} \\
&= \frac{1}{2} \times \frac{1}{2} + \frac{1}{3} \times \frac{1}{2} = \frac{5}{12}.
\end{aligned}$$

The probability of picking a sock is $\frac{5}{12}$.

Note that it is unnecessary to specify $\Omega$ explicitly.

**7.19** If $H_1$ and $H_2$ are disjoint events, then so are $A \cap H_1$ and $A \cap H_2$. It follows that

$$\begin{aligned} P(A) &= P(A \cap (H_1 \cup H)) && \text{since } H_1 \cup H_2 = \Omega \\ &= P((A \cap H_1) \cup (A \cap H_2)) \\ &= P(A \cap H_1) + P(A \cap H_2) && \text{by additivity} \\ &= P(A|H_1)P(H_1) + P(A|H_2)P(H_2) && \text{by Definition 7.2.} \end{aligned}$$

**7.20** If $H_1, \ldots, H_n$ are pairwise disjoint events, then so are $A \cap H_1, \ldots, A \cap H_n$. It follows that

$$\begin{aligned} P(A) &= P(A \cap (H_1 \cup \cdots \cup H_n)) && \text{since } H_1 \cup \cdots \cup H_n = \Omega \\ &= P((A \cap H_1) \cup \cdots \cup (A \cap H_n)) \\ &= P(A \cap H_1) + \cdots + P(A \cap H_n) && \text{by additivity} \\ &= P(A|H_1)P(H_1) + \cdots + P(A|H_n)P(H_n) && \text{by Definition 7.2.} \end{aligned}$$

**7.21** If $H_1, H_2, \ldots$ are pairwise disjoint, then so are the events $A \cap H_1, A \cap H_2, \ldots$ . Thus,

$$\begin{aligned} P(A) &= P\left(A \cap \bigcup_{n=1}^{\infty} H_n\right) && \text{since } \bigcup_{n=1}^{\infty} H_n = \Omega \\ &= P\left(\bigcup_{n=1}^{\infty} (A \cap H_n)\right) \\ &= \sum_{n=1}^{\infty} P(A \cap H_n) && \text{by countable additivity} \\ &= \sum_{n=1}^{\infty} P(A|H_n)P(H_n) && \text{by Definition 7.2.} \end{aligned}$$

**7.22** Consider the following events:

$$\begin{aligned} S &= \text{the person questioned is a bathroom singer,} \\ Y &= \text{the answer is YES,} \\ D_1 &= \text{the die shows 1,} \\ D_{2345} &= \text{the die shows 2, 3, 4, or 5,} \\ D_6 &= \text{the die shows 6.} \end{aligned}$$

According to the rules of the survey,

$$P(Y|D_1) = 0, \quad P(Y|D_{2345}) = P(S), \quad P(Y|D_6) = 1.$$

The events $D_1, D_{2345}, D_6$ are pairwise disjoint and, since one of the outcomes $1, 2, 3, 4, 5, 6$ must occur, the union of $D_1, D_{2345}, D_6$ is the whole probability space (which we do not need to specify explicitly). Therefore, the total probability formula can be applied:

$$\begin{aligned} P(Y) &= P(Y|D_1)P(D_1) + P(Y|D_{2345})P(D_{2345}) + P(Y|D_6)P(D_6) \\ &= 0 \times \frac{1}{6} + P(S) \times \frac{4}{6} + 1 \times \frac{1}{6} \\ &= \frac{1}{6} + \frac{2}{3}P(S). \end{aligned}$$

If $P(Y) = \frac{2}{3}$, we find that $P(S) = \frac{3}{4}$. The probability of being a bathroom singer is $\frac{3}{4}$.

**7.23** Let us consider the events

$$\begin{aligned} W &= \text{the player with the white die wins the game,} \\ R_i &= \text{the red die shows } i, \text{ where } i = 1, \dots, 6. \end{aligned}$$

By the total probability formula,

$$P(W) = P(W|R_1)P(R_1) + \cdots + P(W|R_6)P(R_6).$$

For the red die, which is loaded, we have

$$P(R_6) = \frac{1}{3}, \quad P(R_1) = \cdots = P(R_5) = \frac{2}{15}.$$

To find the conditional probabilities $P(W|R_i)$, observe that if the red dies shows $i$, then the outcomes $i, i+1, \dots, 6$ win the game for the player with the white die. Because the white die is fair,

$$P(W|R_i) = \frac{\#\{i, i+1, \dots, 6\}}{\#\{1, 2, \dots, 6\}} = \frac{6-i+1}{6}.$$

It follows that

$$\begin{aligned} P(W) = \frac{6}{6} \times \frac{2}{15} + \frac{5}{6} \times \frac{2}{15} + \frac{4}{6} \times \frac{2}{15} \\ + \frac{3}{6} \times \frac{2}{15} + \frac{2}{6} \times \frac{2}{15} + \frac{1}{6} \times \frac{1}{3} = \frac{1}{2}. \end{aligned}$$

The probability of winning is $\frac{1}{2}$. In this game, it does not matter which die you choose! The rule that the white die wins in case of equal results is compensated by the fact that the red die is loaded.

**7.24** Consider the same events $D_1$, $D_2$, and $S$ as in Problem 7.18. We have

$$P(D_1) = P(D_2) = \frac{1}{2}, \quad P(S|D_1) = \frac{1}{2}, \quad P(S|D_2) = \frac{1}{3}.$$

Then, the desired probability $P(D_1|S)$ can be found as follows:

$$\begin{aligned} P(D_1|S) &= \frac{P(S \cap D_1)}{P(S)} \\ &= \frac{P(S|D_1)P(D_1)}{P(S|D_1)P(D_1) + P(S|D_2)P(D_2)} \\ &= = \frac{\frac{1}{2} \times \frac{1}{2}}{\frac{1}{2} \times \frac{1}{2} + \frac{1}{3} \times \frac{1}{2}} = \frac{3}{5}. \end{aligned}$$

In the numerator, we used the definition of conditional probability and in the denominator, the total probability formula.

The probability that the sock comes from the first drawer is $\frac{3}{5}$.

**7.25** By the definition of conditional probability and the total probability formula,

$$\begin{aligned} P(H_1|A) &= \frac{P(H_1 \cap A)}{P(A)} \\ &= \frac{P(A|H_1)P(H_1)}{P(A|H_1)P(H_1) + P(A|H_2)P(H_2)}. \end{aligned}$$

**7.26** Using the definition of conditional probability and the total probability formula in the case of $n$ events $H_1, \ldots, H_n$, we obtain

$$\begin{aligned} P(H_1|A) &= \frac{P(H_1 \cap A)}{P(A)} \\ &= \frac{P(A|H_1)P(H_1)}{P(A|H_1)P(H_1) + \cdots + P(A|H_n)P(H_n)}. \end{aligned}$$

**7.27** By Definition 7.2 and Theorem 7.1,

$$\begin{aligned} P(H_1|A) &= \frac{P(H_1 \cap A)}{P(A)} \\ &= \frac{P(A|H_1)P(H_1)}{\sum_{n=1}^{\infty} P(A|H_n)P(H_n)}. \end{aligned}$$

**7.28** Consider the events $F$, $U$, and $H$ defined in the hint. Then,

$$P(F) = P(U) = \frac{1}{2}, \quad P(H|F) = \frac{1}{2}, \quad P(H|U) = \frac{1}{3}.$$

By the Bayes formula,

$$P(F|H) = \frac{P(H|F)P(F)}{P(H|F)P(F) + P(H|U)P(U)} = \frac{\frac{1}{2} \times \frac{1}{2}}{\frac{1}{2} \times \frac{1}{2} + \frac{1}{3} \times \frac{1}{2}} = \frac{3}{5}.$$

The probability that the fair coin has been selected given that it lands heads up is $\frac{3}{5}$.

**7.29** The solution is just as in Problem 7.28, but now

$$P(F) = \frac{2}{3}, \quad P(U) = \frac{1}{3}, \quad P(H|F) = \frac{1}{2}, \quad P(H|U) = p.$$

By the Bayes formula,

$$\begin{aligned} P(F|H) &= \frac{P(H|F)P(F)}{P(H|F)P(F) + P(H|U)P(U)} \\ &= \frac{\frac{1}{2} \times \frac{2}{3}}{\frac{1}{2} \times \frac{2}{3} + p \times \frac{1}{3}} = \frac{1}{1+p}. \end{aligned}$$

The probability that the selected coin is fair given that it lands heads up is $\frac{1}{1+p}$.

**7.30** We shall use the events $S$ and $Y$ defined in the solution to Problem 7.22.

If you are a bathroom singer, your answer will be YES if the die shows 2, 3, 4, 5, or 6; that is,

$$P(Y|S) = \frac{5}{6}.$$

If you are not a bathroom singer, you will say YES only if the die shows 6, so

$$P(Y|\Omega \setminus S) = \frac{1}{6}.$$

From the Bayes formula,

$$\begin{aligned} P(S|Y) &= \frac{P(Y|S)P(S)}{P(Y|S)P(S) + P(Y|\Omega \setminus S)P(\Omega \setminus S)} \\ &= \frac{\frac{5}{6} \times \frac{3}{4}}{\frac{5}{6} \times \frac{3}{4} + \frac{1}{6} \times (1 - \frac{3}{4})} = \frac{15}{16}. \end{aligned}$$

In a similar way, we obtain

$$P(N|S) = \frac{1}{6}, \quad P(N|\Omega \setminus S) = \frac{5}{6}$$

and

$$\begin{aligned} P(S|N) &= \frac{P(N|S)P(S)}{P(N|S)P(S) + P(N|\Omega \setminus S)P(\Omega \setminus S)} \\ &= \frac{\frac{1}{6} \times \frac{3}{4}}{\frac{1}{6} \times \frac{3}{4} + \frac{5}{6} \times (1 - \frac{3}{4})} = \frac{3}{8}. \end{aligned}$$

The odds that a person who answered YES to the survey question is indeed a bathroom singer are $\frac{15}{16}$. For a person who answered NO, the odds are $\frac{3}{8}$.

**7.31** We choose $\Omega$ to be the deck of cards, with uniform probability. Then,

$$P(A) = \frac{\#A}{\#\Omega} = \frac{4}{52} = \frac{1}{13}$$

and

$$P(A|B) = \frac{\#(A \cap B)}{\#B} = \frac{1}{13},$$

so, indeed, $P(A) = P(A|B)$.

**7.32** Using the definition of conditional probability, we can write conditions a) and b) as

$$\text{a) } P(A) = \frac{P(A \cap B)}{P(B)}, \qquad \text{b) } P(B) = \frac{P(B \cap A)}{P(A)}.$$

On multiplying both sides respectively by $P(B)$ and $P(A)$, we can see that each of these conditions is equivalent to c).

**7.33** Using the definition of conditional probability and Problems 4.10 and 4.11, we can write

$$\begin{aligned} P(A|\Omega \setminus B) &= \frac{P(A \cap (\Omega \setminus B))}{P(\Omega \setminus B)} = \frac{P(A \setminus B)}{P(\Omega \setminus B)} \\ &= \frac{P(A) - P(A \cap B)}{1 - P(B)}. \end{aligned}$$

On the other hand,

$$P(A|B) = \frac{P(A \cap B)}{P(B)}.$$

Substituting these expressions for $P(A|\Omega \setminus B)$ and $P(A|B)$, we can write conditions a), b), and c) as

$$\begin{aligned} &\text{a)} \quad P(A) = \frac{P(A \cap B)}{P(B)} \\ &\text{b)} \quad P(A) = \frac{P(A) - P(A \cap B)}{1 - P(B)} \\ &\text{c)} \quad \frac{P(A \cap B)}{P(B)} = \frac{P(A) - P(A \cap B)}{1 - P(B)}. \end{aligned}$$

Now, some elementary algebra shows that each of these conditions is equivalent to $P(A \cap B) = P(A)P(B)$, so they must be equivalent to each other.

**7.34** The equivalence b) $\Leftrightarrow$ c) follows from a) $\Leftrightarrow$ b), so we only need to prove the latter.

Condition b) can be written as

$$P((\Omega \setminus A) \cap B) = P(\Omega \setminus A)P(B).$$

The left-hand side is equal to $P(B \setminus (A \cap B)) = P(B) - P(A \cap B)$ and the right-hand side to $(1 - P(A))\,P(B) = P(B) - P(A)P(B)$, so b) is equivalent to

$$P(A \cap B) = P(A)P(B),$$

i.e., to condition a).

**7.35** $\emptyset, \Omega, \{1,3\}, \{1,4\}, \{2,3\}, \{2,4\}$.

**7.36** Suppose that $B$ has $j$ elements and $A \cap B$ has $m$ elements. Since $A$ and $B$ are independent, $P(A)P(B) = P(A \cap B)$. Because $A$ has 4 elements and $\Omega$ has 6, we obtain $\frac{4}{6}\frac{j}{6} = \frac{m}{6}$, so

$$4j = 6m.$$

This is possible only if $j$ is divisible by 3, i.e., if $j = 0$, 3, or 6.

**7.37** Since $A$ and $B$ are independent, $P(A)P(B) = P(A \cap B)$. Denoting the number of elements of $A \cap B$ by $m$, we obtain $\frac{i}{n}\frac{j}{n} = \frac{m}{n}$, so

$$ij = mn.$$

Divide both sides by $\gcd(i,n)$:

$$j\frac{i}{\gcd(i,n)} = m\frac{n}{\gcd(i,n)}.$$

Because $\frac{i}{\gcd(i,n)}$ and $\frac{n}{\gcd(i,n)}$ have no common divisors other than 1, it follows that

$$j = k\frac{n}{\gcd(i,n)}$$

for some integer $k$. But $0 \leqslant j \leqslant n$, so $k \in \{0, 1, \ldots, \gcd(i,n)\}$.

**7.38** By Definition 7.3,

$$P(A)P(A) = P(A \cap A) = P(A),$$

so $P(A)\,(P(A) - 1) = 0$, which means that $P(A) = 0$ or 1.

**7.39** Condition a) means, in particular, that $A$ is independent of itself, so the implication a) $\Rightarrow$ b) follows from Problem 7.38.

To verify the implication b) $\Rightarrow$ a), observe that

$$\begin{aligned}&\text{if } P(A)=0, \text{ then } P(A\cap B)=0,\\&\text{if } P(A)=1, \text{ then } P(A\cap B)=P(B)\end{aligned}$$

for any event $B$. In the former case, this is because $A\cap B\subset A$, so $P(A\cap B)\leqslant P(A)=0$. In the latter case, $P(B)=P(B\cap A)+P(B\backslash A)$ and $P(B\backslash A)=0$ because $B\backslash A\subset\Omega\backslash A$ and $P(\Omega\backslash A)=1-P(A)=0$.

In both cases, the equality $P(A\cap B)=P(A)P(B)$ is clearly satisfied, i.e., $A$ and $B$ are independent.

**7.40** Each time the coin is tossed, it can land heads or tails up. Hence, if the coin is tossed three times, there are $2\times 2\times 2=8$ possible outcomes, which are equally likely for a fair coin.

Let us count the elements in $A$. There are four of them: HTH, HTT, THH, THT. Similarly, there are four elements in $B$ and $C$, so

$$P(A)=P(B)=P(C)=\frac{4}{8}=\frac{1}{2}.$$

Next, we find the elements of $A\cap B$. There are two: HTH and THT. There are also two elements in $B\cap C$ and $C\cap A$, so

$$P(A\cap B)=P(B\cap C)=P(C\cap A)=\frac{2}{8}=\frac{1}{4}.$$

It follows that

$$\begin{aligned}P(A|B)&=\frac{P(A\cap B)}{P(B)}=\frac{1/4}{1/2}=\frac{1}{2}=P(A),\\P(A|C)&=\frac{P(A\cap C)}{P(C)}=\frac{1/4}{1/2}=\frac{1}{2}=P(A).\end{aligned}$$

Finally, we observe that $A\cap B\cap C$ is empty, so

$$P(A|B\cap C)=\frac{P(A\cap B\cap C)}{P(B\cap C)}=0\neq P(A).$$

**7.41** In Solution 7.40, we computed

$$P(A)=P(B)=P(C)=\frac{1}{2}$$

and

$$P(A\cap B)=P(B\cap C)=P(C\cap A)=\frac{1}{4}.$$

This implies that

$$P(A \cap B) = P(A)P(B),$$

so $A$ and $B$ are independent events by Definition 7.3. Similarly, it follows that

$$P(B \cap C) = P(B)P(C), \quad P(C \cap A) = P(C)P(A),$$

i.e., the events $B$ and $C$ are also independent, and so are $C$ and $A$.

**7.42** Condition a) can be written as

$$P(A)P(B \cap C) = P(A \cap B \cap C).$$

But $P(B \cap C) = P(B)P(C)$ because $B$ and $C$ are independent events, so the above is equivalent to

$$P(A)P(B)P(C) = P(A \cap B \cap C).$$

**7.43** a) Since

$$\begin{aligned} P(A \cap (B \cap C)) &= P(A \cap B \cap C) \\ &= P(A)P(B)P(C) \\ &= P(A)P(B \cap C), \end{aligned}$$

$A$ and $B \cap C$ are independent by Definition 7.3.
b) We use Problem 4.18 to obtain

$$\begin{aligned} P(A \cap (B \cup C)) &= P((A \cap B) \cup (A \cap C)) \\ &= P(A \cap B) + P(A \cap C) - P(A \cap B \cap C) \\ &= P(A)P(B) + P(A)P(C) - P(A)P(B)P(C) \\ &= P(A)\left(P(B) + P(C) - P(B)P(C)\right) \\ &= P(A)P(B \cup C), \end{aligned}$$

so $A$ and $B \cup C$ are independent by Definition 7.3.
c) Using Problem 4.10, we obtain

$$\begin{aligned} P(A \cap (B \setminus C)) &= P((A \cap B) \setminus (A \cap B \cap C)) \\ &= P(A \cap B) - P(A \cap B \cap C) \\ &= P(A)P(B) - P(A)P(B)P(C) \\ &= P(A)\left(P(B) - P(B)P(C)\right) \\ &= P(A)\left(P(B) - P(B \cap C)\right) \\ &= P(A)P(B \setminus C). \end{aligned}$$

Again, by Definition 7.3, the events $A$ and $B \setminus C$ are independent.

**7.44** The $\sigma$-field $\mathcal{F}_{\{B,C\}}$ generated by $\{B, C\}$ is, in fact, a finite field with at most four atoms: $B \cap C$, $B \setminus C$, $C \setminus B$, and $\Omega \setminus (B \cup C)$ (some of these sets my be empty and then should be discarded). Each event $D \in \mathcal{F}_{\{B,C\}}$ is a finite union of some of these atoms, so it can be expressed by means of finitely many unions, intersections, and complements of $B$ and $C$. It follows by Problem 7.43 that $A$ and $D$ are independent.

**7.45** We shall prove the equivalence a) $\Leftrightarrow$ b). Then, a) $\Leftrightarrow$ c) and a) $\Leftrightarrow$ d) will follow by applying a) $\Leftrightarrow$ b) repeatedly.

First, we observe that by Problem 7.34, $A$, $B$, and $C$ are pairwise independent if and only if $\Omega \setminus A$, $B$, and $C$ are. Now, consider the condition

$$P((\Omega \setminus A) \cap B \cap C) = P(\Omega \setminus A)P(B)P(C).$$

The left-hand side is equal to

$$\begin{aligned} P((\Omega \setminus A) \cap B \cap C) &= P((B \cap C) \setminus (A \cap B \cap C)) \\ &= P(B \cap C) - P(A \cap B \cap C) \\ &= P(B)P(C) - P(A \cap B \cap C), \end{aligned}$$

whereas the right-hand side is equal to

$$\begin{aligned} P(\Omega \setminus A)P(B)P(C) &= (1 - P(A))P(B)P(C) \\ &= P(B)P(C) - P(A \cap B \cap C), \end{aligned}$$

so the above condition is equivalent to

$$P(A \cap B \cap C) = P(A)P(B)P(C),$$

completing the proof.

**7.46** Denote by $\mathcal{F}$ the family of events $C$ such that $B$ and $C$ are independent. We claim that $\mathcal{F}$ is a $\sigma$-field. First, $\Omega \in \mathcal{F}$ by Problem 7.39. Second, if $C \in \mathcal{F}$, then $\Omega \setminus C \in \mathcal{F}$ by Problem 7.34. Third, for any sequence of events $C_1, C_2, \ldots \in \mathcal{F}$

$$\begin{aligned} P(B \cap (C_1 \cup C_2 \cup \cdots)) &= P((B \cap C_1) \cup (B \cap C_2) \cup \cdots) \\ &= P(B \cap C_1) + P(B \cap C_2) + \cdots \\ &= P(B)P(C_1) + P(B)P(C_2) + \cdots \\ &= P(B)(P(C_1) + P(C_2) + \cdots) \\ &= P(B)P(C_1 \cup C_2 \cup \cdots) \end{aligned}$$

if the $C_i$ are pairwise disjoint, which can be assumed without loss of generality. It follows that $C_1 \cup C_2 \cup \cdots \in \mathcal{F}$. We have verified that $\mathcal{F}$ is a $\sigma$-field.

Now, since $A_1, \ldots, A_n \in \mathcal{F}$, it follows that $\mathcal{F}_{\{A_1,\ldots,A_n\}} \subset \mathcal{F}$, completing the proof.

**7.47** Let $\mathcal{F}$ be the family of events $D$ such that $C$ and $D$ are independent for any $C$ in $\mathcal{F}_{\{A_1,\dots,A_n\}}$. In a similar way as in Solution 7.46, we can verify that $\mathcal{F}$ is a $\sigma$-field. Since $B_1, \dots, B_m \in \mathcal{F}$, it follows that $\mathcal{F}_{\{B_1,\dots,B_m\}} \subset \mathcal{F}$, completing the proof.

# 8

# Random Variables and Their Distributions

## 8.1 Theory and Problems

Suppose that you are invited to play a game with the following rules: One of the numbers $2, \dots, 12$ is chosen at random by throwing a pair of dice and adding the numbers shown. You win 9 dollars in case 2, 3, 11, or 12 comes out, or lose 10 dollars if the outcome is 7. Otherwise, you do not win or lose anything. This defines a function on the set of all possible outcomes $\{2, \dots, 12\}$, the value of the function being the corresponding gain (or loss if the value is negative). You may want to know, for example, the probability that the function takes positive values, i.e., that you will win some money.

In general, for a function $X : \Omega \to \mathbb{R}$ defined on a probability space $\Omega$ with $\sigma$-field $\mathcal{F}$ and probability measure $P$, we can talk of the probability that $a < X$ for some real number $a$ only if the set $\{\omega \in \Omega : a < X(\omega)\}$ is an event, i.e., it belongs to $\mathcal{F}$. This gives rise to the following definition.

► **Definition 8.1** We call $X : \Omega \longrightarrow \mathbb{R}$ a *random variable* if

$$\{\omega \in \Omega : a < X(\omega)\} \in \mathcal{F}$$

for every $a \in \mathbb{R}$.

This terminology is rather unfortunate. In fact, $X$ is not a variable at all, but a function. Nevertheless, the term "random variable" is traditional and generally accepted.

Let us recall that the *inverse image* $\{X \in B\}$ (sometimes also denoted by $X^{-1}(B)$) of a set $B \subset \mathbb{R}$ under a function $X : \Omega \to \mathbb{R}$ is defined by

$$\{X \in B\} = \{\omega \in \Omega : X(\omega) \in B\}.$$

For brevity, we shall write $\{X > a\}$ instead of $\{X \in (a, \infty)\}$, $\{X \leqslant a\}$ instead of $\{X \in (-\infty, a]\}$, $\{X = a\}$ instead of $\{X \in \{a\}\}$, and the like.

The following problem provides an equivalent condition for $X$ to be a random variable.

**8.1** Show that $X : \Omega \to \mathbb{R}$ is a random variable if and only if the inverse image $\{X \in B\}$ belongs to $\mathcal{F}$ for each Borel set $B \subset \mathbb{R}$.

Now we are in a position to ask more general questions than before: If $B$ is a given Borel set, what is the probability that the values of $X$ are in $B$?

**8.2** If $X$ denotes the amount won (or lost) in the game described at the beginning of this chapter, what are the probabilities $P(\{X > 0\})$ and $P(\{X < 0\})$?

**8.3** Two dice are rolled. Let $X$ be the larger of the two numbers shown. Compute $P(\{X \in [2, 4]\})$.

**8.4** Let $\Omega = [0, 1]$ with Borel sets and Lebesgue measure. Find $P(\{X \in [0, \frac{1}{2})\})$ for $X(x) = x^2$.

**8.5** Let $X$ be the number of tosses of a fair coin up to and including the first toss showing heads. Find $P(\{X \in 2\mathbb{N}\})$, where $2\mathbb{N} = \{2n : n = 1, 2, \ldots\}$ is the set of even integers.

It is clear from Definition 8.1 and Problem 8.1 that whether a function $X : \Omega \to \mathbb{R}$ is a random variable or not depends on the choice of the $\sigma$-field $\mathcal{F}$ on $\Omega$.

**8.6** Show that if $X$ is a constant function, then it is a random variable with respect to any $\sigma$-field.

**8.7** Let $\Omega = \{1, 2, 3, 4\}$ and $\mathcal{F} = \{\varnothing, \Omega, \{1\}, \{2, 3, 4\}\}$. Is $X(\omega) = 1 + \omega$ a random variable with respect to the $\sigma$-field $\mathcal{F}$? If not, give an example of a non-constant function which is.

**8.8** For each of the functions below, find the smallest $\sigma$-field on $\Omega = \{-2, -1, 0, 1, 2\}$ with respect to which the function is a random variable:

a) $X(\omega) = \omega^2$,
b) $X(\omega) = \omega + 1$.

∘ **8.9** Find the smallest $\sigma$-fields on $\Omega = \{-2, -1, 0, 1, 2\}$ such that the following functions are random variables:
a) $X(\omega) = |\omega|$,
b) $X(\omega) = 2\omega$.

**8.10** Consider $\Omega = [0, 1]$ with the $\sigma$-field $\mathcal{F}$ of Borel sets $B$ contained in $[0, 1]$.
a) Is $X(x) = x$ a random variable on $\Omega$ with respect to $\mathcal{F}$?
b) Is $Y(x) = |x - \frac{1}{2}|$ a random variable on $\Omega$ with respect to $\mathcal{F}$?

**8.11** Let $\Omega = [0, 1]$ with the $\sigma$-field $\mathcal{G}$ of Borel sets $B$ contained in $[0, 1]$ such that $B = 1 - B$. (By $1 - B$ we denote the set $\{1 - x : x \in B\}$.)
a) Is $X(x) = x$ a random variable on $\Omega$ with respect to $\mathcal{G}$?
b) Is $Y(x) = |x - \frac{1}{2}|$ a random variable on $\Omega$ with respect to $\mathcal{G}$?

**8.12** What is the smallest number of elements of a $\sigma$-field if a function $X : \Omega \to \mathbb{R}$ taking exactly $n$ different values is to be a random variable with respect to this $\sigma$-field?

**8.13** Show that if $X$ is a random variable with respect to a $\sigma$-field $\mathcal{F}$ and $\mathcal{F} \subset \mathcal{G}$ for some $\sigma$-field $\mathcal{G}$, then $X$ is a random variable with respect to $\mathcal{G}$.

In many situations, a random variable $X$ may be specified only to within defining the probabilities $P(\{X \in B\})$ for all Borel sets $B$ in $\mathbb{R}$, whereas the underlying probability space $\Omega$ does not even need to be introduced explicitly. (All the same, it is assumed that a suitable probability space exists.) In particular, two random variables $X$ and $X'$ defined on two (possibly) different probability spaces $\Omega$ and $\Omega'$ with $\sigma$-fields $\mathcal{F}$ and $\mathcal{F}'$ and probability measures $P$ and $P'$, respectively, are considered to be equivalent for such purposes as long as

$$P(\{X \in B\}) = P'(\{X' \in B\})$$

for all Borel sets $B \subset \mathbb{R}$. The map assigning $P(\{X \in B\})$ to any Borel set $B \subset \mathbb{R}$ turns out to be a probability measure.

**8.14** Let $X$ be a random variable. Show that the map

$$P_X : B \longmapsto P(\{X \in B\})$$

is a probability measure on the $\sigma$-field $\mathcal{B}$ of Borel sets in $\mathbb{R}$.

▶ **Definition 8.2** We call the probability measure $P_X$ the *distribution* of a random variable $X$. It is a probability measure on $\mathbb{R}$ defined on the $\sigma$-field $\mathcal{B}$ of Borel sets.

**8.15** Find an example of two different random variables $X$ and $X'$ with the same distribution $P_X = P_{X'}$.

Below we have a series of exercises on evaluating the distribution $P_X$ of a given random variable $X$.

**8.16** Let $X$ be the amount won (or lost) in the game described at the beginning of this chapter. Find $P_X((0, +\infty))$ and $P_X((-\infty, 0))$.

**8.17** Two dice are rolled. Let $X$ be the larger of the two numbers shown. Compute $P_X([2, 4])$.

**8.18** Let $\Omega = [0, 1]$ with Borel sets and Lebesgue measure. Find $P_X([0, \frac{1}{2}))$ if $X(x) = x^2$.

**8.19** Let $X$ be the number of tosses of a fair coin up to and including the first toss showing heads. Find $P_X(2\mathbb{N})$, where $2\mathbb{N} = \{2n : n = 1, 2, \ldots\}$ is the set of even integers.

It proves convenient to describe the distribution of a random variable by means of a real-valued function defined as follows.

▶ **Definition 8.3** The function $F_X : \mathbb{R} \to \mathbb{R}$ defined by

$$F_X(x) = P_X((-\infty, x])$$

for all $x \in \mathbb{R}$ is called the *distribution function* of a random variable $X$.

Find the distribution function of each of the random variables below.

**8.20** $X(\omega) \equiv c$ (a constant random variable identically equal to $c$).

**8.21** $X(\omega) = 1$ for $\omega \in A$ and $X(\omega) = 2$ otherwise, where $P(A) = \frac{1}{3}$.

**8.22** $X(\omega) = c_k$ with probability $\alpha_k$ for $k = 1, \ldots, n$, where $c_1 < c_2 < \cdots < c_n$ and $\alpha_1 + \cdots + \alpha_n = 1$.

∘ **8.23** $X(\omega) = 1$ for $\omega \in A$, $X(\omega) = -2$ for $\omega \in B$, and $X(\omega) = 2$ otherwise, where $P(A) = \frac{1}{3}$, $P(B) = \frac{1}{2}$, and $A \cap B = \emptyset$.

**8.24** $\Omega = [0, 1]$ with Lebesgue measure, $X(x) = 2x - 1$.

**8.25** Let $X$ be the number of tosses of a fair coin up to and including the first toss showing heads. Compute and sketch the distribution function $F_X$.

Verify the following properties of the distribution function.

**8.26** $F_X$ is non-decreasing, i.e., if $y_1 \leqslant y_2$, then $F_X(y_1) \leqslant F_X(y_2)$.

**8.27** $\lim_{y\to+\infty} F_X(y) = 1$.

**8.28** $\lim_{y\to-\infty} F_X(y) = 0$.

**8.29** Show that if all the values of $X$ are in an interval $[a,b]$, where $a < b$, then $F_X(y) = 0$ for all $y < a$ and $F_X(y) = 1$ for all $y \geqslant b$.

$*$ **8.30** $F_X$ is right-continuous.

**8.31** $F_X$ is continuous at $y$ if and only if $P_X(\{y\}) = 0$.

**8.32** Let $a < b$. Find expressions for the following probabilities in terms of $F_X$:

a) $P_X((a,b])$, c) $P_X([a,b])$,
b) $P_X((a,b))$, d) $P_X([a,b))$.

Below there are some exercises on finding the distribution function.

**8.33** Find $F_X$ if $\Omega = [0,1]\times[0,1]$ is the unit square with uniform measure (which measures the surface area) and $X(\omega)$ is the distance between $\omega \in \Omega$ and the nearest edge of the square.

**8.34** A coin is tossed. If it shows heads, you pay 2 dollars. If it shows tails, you spin a wheel which gives the amount you win distributed with uniform probability between 0 and 10 dollars. Your gain (or loss) is a random variable $X$. Find the distribution function and use it to compute the probability that you will not win at least 5 dollars.

**8.35** Your car has broken down at noon. You have called a repair team, which left at 12:00 from a place 100 miles from where you are. They travel at 100 miles per hour. In the meantime, you try to fix the car yourself. The chances of success are uniform in the time interval between 12 and 1 o'clock with total probability $\frac{1}{2}$. If you succeed, you will drive to meet the repair team at 100 miles per hour. Let $X$ be the distance traveled by the team before they meet you (the fee you pay depends on it). Find $F_X$.

**8.36** Two towns A and B are 50 miles apart. Car 1 leaves town A and car 2 leaves town B independently of one another. The departure time of each car is random and uniformly distributed between 12 and 1 o'clock. The speed of each car is 100 miles per hour and they travel directly to meet one another. Denote by $X$ the distance between the meeting point and A. Find $F_X$.

Next, we shall consider some important particular cases, where the distribution of a random variable is of a special form.

▶ **Definition 8.4** A random variable $X$ is called *discrete* whenever there is a countable set $C \subset \mathbb{R}$ such that $P_X(C) = 1$.

**8.37** Show that $X$ is a discrete random variable if and only if the distribution $P_X$ is a discrete probability measure on $\mathbb{R}$ in the sense of Definition 6.7.

Below we shall consider some typical examples of discrete random variables and discrete distributions.

**8.38** A random variable $X$ is said to have the *binomial distribution* $B(n,p)$, where $n = 1, 2, \ldots$ and $p \in [0,1]$, if

$$P_X(\{k\}) = \binom{n}{k} p^k (1-p)^{n-k} \tag{8.1}$$

for $k = 0, 1, \ldots, n$. Verify that such a random variable is discrete.

**8.39** Show that the indicator function $\mathbb{I}_A$ of an event $A$ has the binomial distribution $B(n,p)$ with $n = 1$ and $p = P(A)$. (This is sometimes called the *Bernoulli distribution.*)

**8.40** Let $X$ have the binomial distribution $B(n,p)$. Show that for any $k = 1, \ldots, n$,

$$P_X(\{k\}) = \frac{p}{1-p} \frac{n-k+1}{k} P_X(\{k-1\}).$$

(This formula can be used in practice to compute $P_X(\{k\})$ by induction starting from $P_X(\{0\}) = (1-p)^n$ or by backward induction starting from $P_X(\{n\}) = p^n$.)

**8.41** Let $X$ have the binomial distribution $B(n,p)$ and let $Y$ have the binomial distribution $B(n, 1-p)$. Verify that for any $k = 0, 1, \ldots, n$,

$$P_X(\{k\}) = P_Y(\{n-k\}).$$

**8.42** A biased coin with probability of heads $p$ and tails $1-p$ is flipped $n$ times. Let $X$ denote the number of heads obtained. What is the distribution of $X$?

**8.43** A biased coin with probability of heads $p$ and tails $1-p$ is flipped until $k$ heads are obtained.

a) Find the distribution of the number of flips $Y$. (This is called the *negative binomial distribution.*)

b) Show that $Y$ is a discrete random variable.

**8.44** Let $X_n$ be a sequence of random variables with binomial distribution $B(n, \frac{\lambda}{n})$ for some $\lambda > 0$. Show that

$$P(\{X_n = k\}) \to e^{-\lambda} \frac{\lambda^k}{k!} \quad \text{as } n \to \infty.$$

The limit in Problem 8.44 defines another well-known distribution considered below.

**8.45** A random variable $X$ is said to have the *Poisson distribution* with parameter $\lambda > 0$ if

$$P_X(\{k\}) = e^{-\lambda} \frac{\lambda^k}{k!} \tag{8.2}$$

for $k = 0, 1, 2, \ldots$ . Show that $X$ is a discrete random variable.

**8.46** A safety device in a laboratory is set to activate an alarm if it registers five or more radioactive particles within one second. If the background radiation is such that the number of particles reaching the device has the Poisson distribution with parameter $\lambda = 0.5$, how likely is it that the alarm will be activated within a given 1-second period?

**8.47** Let $X$ have the Poisson distribution with parameter $\lambda$. Verify that

$$P_X(\{k+1\}) = \frac{\lambda}{k+1} P_X(\{k\})$$

for each $k = 0, 1, 2, \ldots$ . (This formula can be used to compute $P_X(\{k\})$ by induction starting from $P_X(\{0\}) = e^{-\lambda}$.)

▶ **Definition 8.5** A random variable $X$ is said to be *absolutely continuous* with *density* $f_X : \mathbb{R} \to [0, \infty)$ if

$$P_X((a, b]) = \int_a^b f_X(x)\, \mathrm{d}x$$

for every $a, b \in \mathbb{R}$. In other words, $P_X$ is an absolutely continuous probability measure on $\mathbb{R}$ with density $f_X$ (see Definition 6.5).

Let $X$ be an absolutely continuous random variable with density $f_X$.

**8.48** Show that $\int_{-\infty}^{\infty} f_X(x)\, \mathrm{d}x = 1$.

**8.49** Show that $P_X(\{y\}) = 0$ for any $y \in \mathbb{R}$.

**8.50** Give a formula for $F_X(y)$ in terms of $f_X$.

**8.51** Show that for any absolutely continuous random variable, the distribution function is continuous.

**8.52** Show that if the distribution function $F_X$ is differentiable with continuous derivative, then $X$ is an absolutely continuous random variable with density $f_X(x) = \frac{dF_X(x)}{dx}$.

**8.53** A random variable $X$ has the *normal distribution* $N(m, \sigma^2)$, where $m, \sigma \in \mathbb{R}$, if it is an absolutely continuous random variable with density

$$f_X(x) = \frac{1}{\sqrt{2\pi\sigma^2}} \exp\left(-\frac{(x-m)^2}{2\sigma^2}\right).$$

Verify that $f_X$ is indeed a density.

**8.54** Let $X$ be a random variable with normal distribution $N(0,1)$. Show that $a + bX$, where $a, b \in \mathbb{R}$, has the normal distribution $N(a, b^2)$.

**8.55** Let $X$ be a random variable with normal distribution $N(0,1)$. Find the density of $Y = e^X$. (The distribution of $Y$ is called the *log normal distribution*.)

**8.56** Let $X$ be an absolutely continuous random variable with uniform distribution on $[0,1]$, i.e., with $f_X(x) = 1$ for any $x \in [0,1]$ and $f_X(x) = 0$ otherwise. Find a function $g : \mathbb{R} \to \mathbb{R}$ such that $g(X)$ has the normal distribution $N(0,1)$.

So far we have studied probability distributions of single random variables. However, there is often a need to look at two or more random variables together. The key notion in this context is the joint distribution. For a single random variable, the distribution is a probability measure on Borel sets in $\mathbb{R}$. In the case of two random variables, we shall consider Borel sets in $\mathbb{R}^2$, i.e., sets belonging to the $\sigma$-field generated by the family of rectangles $(a,b) \times (c,d) \subset \mathbb{R}^2$. To begin with, we need to establish the following property:

**8.57** Suppose that $X$ and $Y$ are random variables defined on the same probability space $(\Omega, \mathcal{F}, P)$. Show that $\{(X,Y) \in B\}$ belongs to $\mathcal{F}$ for any Borel set $B$ in $\mathbb{R}^2$.

This ensures that the set $\{(X,Y) \in B\}$ in the following definition belongs to $\mathcal{F}$ and it makes sense to consider its probability.

▶ **Definition 8.6** The *joint distribution* of random variables $X$ and $Y$ is a probability measure $P_{X,Y}$ on $\mathbb{R}^2$ such that

$$P_{X,Y}(B) = P(\{(X,Y) \in B\})$$

for every Borel set $B \subset \mathbb{R}$.

**8.58** Let $X$ and $Y$ be random variables defined on the same probability space. Show that the distribution $P_X$ of $X$ and $P_Y$ of $Y$ can be obtained from the joint distribution as follows:

$$P_X(B) = P_{X,Y}(B \times \mathbb{R}),$$
$$P_Y(B) = P_{X,Y}(\mathbb{R} \times B)$$

for any Borel set $B$ in $\mathbb{R}$.

When expressed in this way $P_X$ and $P_Y$ are referred to as the *marginal distributions*.

**8.59** Two identical coins are flipped simultaneously. Let $X$ be the number of heads and $Y$ the number of tails shown. What is the joint distribution of $X$ and $Y$? What are the marginal distributions?

**8.60** A pair of dice are rolled. Let $X$ be the greater and $Y$ the smaller of the numbers shown. Find the joint distribution and the marginal distributions of $X$ and $Y$.

**8.61** Two balls are drawn from an urn containing one yellow, two red and three blue balls. Let $X$ be the number of red balls and $Y$ the number of blue balls drawn. Find the joint distribution and the marginal distributions of $X$ and $Y$.

▶ **Definition 8.7** Random variables $X$ and $Y$ are said to be jointly discrete whenever there is a countable set $C \subset \mathbb{R}^2$ such that $P_{X,Y}(C) = 1$.

**8.62** Prove that random variables $X$ and $Y$ are jointly discrete if and only if they are both discrete.

**8.63** A fair coin is tossed repeatedly. Suppose that heads appear for the first time after $X$ tosses and tails appear for the first time after $Y$ tosses. Show that $X$ and $Y$ are jointly discrete random variables and find their joint distribution. Hence or otherwise, compute the marginal distributions of $X$ and $Y$.

**8.64** Let $X$ and $Y$ have the joint distribution

$$P_{X,Y}(\{(m,n)\}) = \begin{cases} \frac{1}{2^{m+1}} & \text{if } m \geqslant n \\ 0 & \text{if } m < n, \end{cases}$$

for $m, n = 1, 2, \ldots$ . Compute the marginal distributions $P_X$ and $P_Y$.

▶ **Definition 8.8** Random variables $X$ and $Y$ are said to be *jointly continuous* if there is a function $f_{X,Y} : \mathbb{R}^2 \to [0, \infty)$, called the *joint density*, such that

$$P_{X,Y}((a,b] \times (c,d]) = \int_c^d \int_a^b f_{X,Y}(x,y)\, dx\, dy$$

for every $a \leqslant b$ and $c \leqslant d$.

**8.65** Show that if $X$ and $Y$ are jointly continuous random variables, then both $X$ and $Y$ are absolutely continuous with density, respectively,

$$f_X(x) = \int_{-\infty}^{+\infty} f_{X,Y}(x,y)\,\mathrm{d}y, \tag{8.3}$$

$$f_Y(y) = \int_{-\infty}^{+\infty} f_{X,Y}(x,y)\,\mathrm{d}x. \tag{8.4}$$

If expressed in this way, $f_X$ and $f_Y$ are called the *marginal densities*.

* **8.66** Find two absolutely continuous random variables $X$ and $Y$ that are not jointly continuous.

**8.67** If $X$ and $Y$ are jointly continuous random variables, show that $U = X+Y$ and $V = X-Y$ are also jointly continuous with joint density

$$f_{U,V}(u,v) = \tfrac{1}{2} f_{X,Y}\left(\tfrac{u+x}{2}, \tfrac{u-v}{2}\right).$$

▶ **Definition 8.9** We say that random variables $X$ and $Y$ are independent whenever the events $\{X \leqslant a\}$ and $\{Y \leqslant b\}$ are independent for any $a, b \in \mathbb{R}$.

**8.68** Show that $X$ and $Y$ are independent random variables if and only if the events $\{X \in A\}$ and $\{Y \in B\}$ are independent for any Borel sets $A, B \subset \mathbb{R}$.

**8.69** Let $X$ and $Y$ be independent random variables. Show that $f(X)$ and $g(Y)$ are independent for any Borel functions $f$ and $g$. (By $f(X)$, we denote the random variable $\Omega \ni \omega \mapsto f(X(\omega)) \in \mathbb{R}$ and by $g(Y)$ the random variable $\Omega \ni \omega \mapsto g(Y(\omega)) \in \mathbb{R}$.)

**8.70** Show that jointly continuous random variables are independent if and only if for every $x, y \in \mathbb{R}$,

$$f_{X,Y}(x,y) = f_X(x) f_Y(y).$$

**8.71** Suppose that $X$ and $Y$ are jointly continuous random variables with joint density $f_{X,Y}(x,y) = ce^{x+y}$ for $x, y \in (-\infty, 0]$ and $f_{X,Y}(x,y) = 0$ otherwise.

a) What is the value of $c$?
b) What is the probability that $X < Y$?
c) What are the marginal densities $f_X$ and $f_Y$?
d) Show that $X$ and $Y$ are independent.

**8.72** Let $X$ and $Y$ be independent random variables with binomial distribution $B(m,p)$ and $B(n,p)$, respectively. What is the distribution of $X+Y$?

**8.73** Let $X$ and $Y$ be independent random variables having the Poisson distribution with parameter $\lambda$ and $\mu$, respectively. Find the distribution of $X + Y$.

**8.74** Suppose that $X$ and $Y$ are jointly continuous independent random variables. Show that the density of the sum $X+Y$ is the *convolution* of the marginal densities:

$$f_{X+Y}(z) = \int_{-\infty}^{+\infty} f_X(x) f_Y(z - x)\, \mathrm{d}z.$$

## 8.2 Hints

**8.1** The "if" part of the statement should be obvious. The "only if" part is based on the fact that the $\sigma$-field of Borel sets is generated by intervals.

**8.2** $P(\{X > 0\})$ is the probability that $X$ is positive, i.e., that you actually win some money. In the case in hand, the only positive value of $X$ is 9. What is the probability that it is achieved?

Similarly, $P(\{X < 0\})$ is the probability of losing the game. What are the negative values that $X$ can take?

**8.3** How many possible outcomes are there if two dice are rolled? How many of these will give $X = 2$? How many will give $X = 3$? How many will give $X = 4$?

**8.4** Sketch the graph of $x^2$ and find the inverse image of $[0, \frac{1}{2})$ in $[0, 1]$.

**8.5** For a given positive integer $n$, what is the probability that $X = n$, i.e., that heads will appear for the first time at the $n$th toss of the coin? Now add up these probabilities for all even positive integers and you will find $P(\{X \in 2\mathbb{N}\})$.

**8.6** The inverse image of a Borel set $B$ depends on whether $c \in B$ or $c \notin B$.

**8.7** Find $\{X \in \{3\}\}$.

**8.8** Either function takes only finitely many values. Find the inverse images of one-element sets containing these values. Then, complete to a $\sigma$-field (in fact, a field).

**8.9** See Problem 8.8.

**8.10** For any Borel set $B \subset \mathbb{R}$, you need to write the inverse images $\{X \in B\}$ and $\{Y \in B\}$ in such a way that it becomes evident whether or not they are Borel sets in $[0,1]$.

**8.11** See Solution 8.10, where the inverse images $\{X \in B\}$ and $\{Y \in B\}$ are computed for any Borel set $B \subset \mathbb{R}$. Apart from showing that these are Borel sets contained in $[0,1]$, you need to verify an additional condition to decide whether or not they belong to $\mathcal{G}$.

**8.12** For an example of such a function with $n = 3$, see Problem 8.8 a). Also refer to Solution 3.26, which relates the number of elements in a finite field of sets to the number of atoms.

**8.13** Does the inverse image of any Borel set $B$ belong to $\mathcal{G}$?

**8.14** Use the fact that $P$ is a probability measure, combined with certain properties of inverse images.

**8.15** It is possible to find examples even if the distribution is just a Dirac measure.

**8.16** See Problem 8.2.

**8.17** See Problem 8.3.

**8.18** See Problem 8.4.

**8.19** See Problem 8.5.

**8.20** To find $F_X(x)$, consider two cases: $x < c$ and $x \geqslant c$.

**8.21** To find $F_X(x)$ consider three cases: $x < 1$, $1 \leqslant x < 2$, and $2 \leqslant x$.

**8.22** Consider the cases when $x < c_1$, $c_{k-1} \leqslant x < c_k$ for $k = 2, \ldots, n$, and $c_n \leqslant x$.

**8.23** See Problems 8.21 and 8.22.

**8.24** Sketch the graph of $X$ and solve the inequality $2x - 1 \leqslant y$ for $x \in [0,1]$.

**8.25** The probabilities $P(\{X = n\})$ for all integers $n$ are given in Solution 8.5.

**8.26** For $y_1 \leqslant y_2$, compare the sets $\{\omega : X(\omega) \leqslant y_1\}$ and $\{\omega : X(\omega) \leqslant y_2\}$ (is one of them contained in the other?) and use Problem 4.12.

**8.27** Consider any sequence $y_n \nearrow \infty$. What can you say about the sequence of sets $\{X \leqslant y_n\}$?

**8.28** Consider any sequence $y_n \searrow -\infty$. What can you say about the sequence of sets $\{X \leqslant y_n\}$?

**8.29** First, suppose that $y < a$. What is the probability that $X \leqslant y$ if all the values of $X$ are greater than or equal to $a$? Next, suppose that $y \geqslant b$. What is the probability that $X \leqslant y$ if none of the values of $X$ exceeds $b$?

**8.30** Consider $y_n \searrow y$ and show that $F_X(y_n) \to F_X(y)$. To do this, draw on the experience gathered in Problem 8.28.

**8.31** By Problem 8.30, we only need to show that $F_X$ is left-continuous at $y$ if and only if $P_X(\{y\}) = 0$. Consider $y_n \nearrow y$ and use an argument similar to that in Solution 8.27.

**8.32** a) Try to write the interval $(a, b]$ in terms of intervals of the form $(-\infty, y]$ and then use the definition of $F_X$ combined with the basic properties of measures.
b) Write the set $(a, b)$ in terms of a sequence of intervals $(a, b_n]$ such that $b_n \nearrow b$.
c) Write $[a, b]$ in terms of a sequence of intervals $(a_n, b]$ such that $a_n \nearrow a$.
d) Write $[a, b)$ by means of intervals of the form $[a, b_n]$, where $b_n \nearrow b$.

**8.33** Look at Figure 8.1, showing the points for which the distance to the nearest edge is less than $x$.

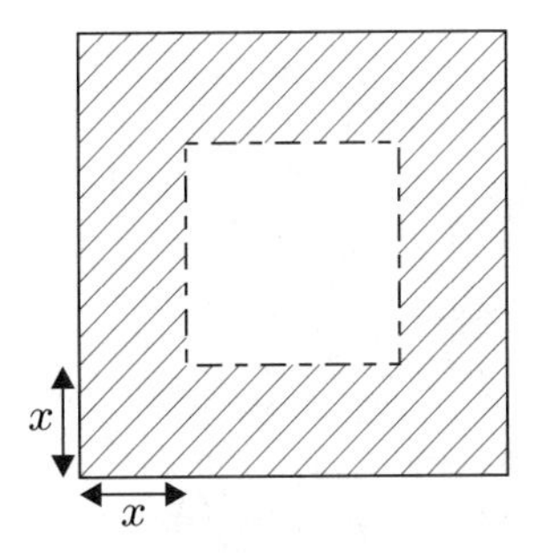


FIGURE 8.1. Distance to the nearest edge less than $x$

**8.34** Consider $F_X(x)$ in the following four cases: $x < -2$, $-2 \leqslant x < 0$, $0 \leqslant x < 10$, and $10 \leqslant x$. Note that the possible values of $X$ fill the set $\{-2\} \cup [0, 10]$. What is the probability that the wheel will show an amount less than $x$ dollars?

**8.35** With probability $\frac{1}{2}$, you will not repair the car at all and the repair team will have to drive 100 miles. So, $P(\{X = 100\}) = \frac{1}{2}$ and the distribution function cannot be continuous. Now, suppose that you have fixed the car in a fraction of an hour equal to $x$ and compute the meeting point. What is the probability that you will meet the team at that point?

**8.36** Consider the sample space $\Omega = [0,1]\times[0,1]$ with uniform probability measure $P$ (one that measures the surface area), where $(t_1, t_2) \in \Omega$ represents the departure times of car 1 and car 2, respectively.

If car 1 leaves at least half an hour earlier than car 2, it will arrive at town B prior to the departure of car 2, so they will meet at B. Which part of $\Omega$ corresponds to this event? What is the value of $X$ on this subset of $\Omega$?

Which subset of $\Omega$ corresponds to the event that the cars will meet at A, i.e., car 2 leaves at least half an hour before car 1? What is the value of $X$ on this subset?

Now, suppose that car 2 leaves no more than half an hour before car 1. Which subset of $\Omega$ corresponds to this event? To find $X$ in this subset, compute the distance traveled by car 1 before meeting the other car.

Finally, consider the case when car 2 leaves no more than half an hour after car 1.

**8.37** Consider the set of all $x \in \mathbb{R}$ such that $P(\{X = x\}) > 0$. This must be a countable set if $X$ is to be a discrete random variable.

**8.38** You will need the *binomial formula*

$$(a+b)^n = \sum_{k=0}^{n} \binom{n}{k} a^k b^{n-k}. \tag{8.5}$$

**8.39** What is the probability that $\mathbb{I}_A = 1$? What is the probability that $\mathbb{I}_A = 0$?

**8.40** This is a simple consequence of (8.1).

**8.41** This also follows immediately from (8.1). The equality $P_X(\{k\}) = P_Y(\{n-k\})$ simply means that for $n$ tosses of a (biased) coin, the probability of $k$ heads is the same as the probability of $n-k$ tails.

**8.42** What is the probability of any given outcome with $k$ heads and $n-k$ tails? Refer to Problem 2.3 to find the number of such outcomes.

**8.43** a) The event $\{Y = n\}$ is the intersection of the independent events "the first $n-1$ flips include $k-1$ heads" and "the $n$th flip is a head."
b) Try to write $P_Y$ as $\sum_{n=k}^{\infty} \alpha_n \delta_{x_n}$ for suitable $x_n$'s and $\alpha_n$'s. You will need to make sure that the $\alpha_n$ add up to 1. The Taylor expansion of $(1-x)^{-k}$ around 0 might come useful.

**8.44** Use (8.1) to express $P(\{X_n = k\})$. Recall that the sequence $\left(1 - \frac{\lambda}{n}\right)^n$ tends to $e^{-\lambda}$ as $n \to \infty$.

**8.45** You will need the Taylor expansion $\sum_{k=0}^{\infty} \frac{x^k}{k!}$ of $e^x$ around 0.

**8.46** What is the probability that the alarm will *not* be activated?

**8.47** This follows immediately from (8.2).

**8.48** Consider $\int_{-n}^{n} f_X(x)\, dx$ and then pass to the limit as $n \to \infty$.

**8.49** Intuitively, to find $P_X(\{y\})$, we integrate $f_X$ from $y$ to $y$, which gives zero.

**8.50** First, express $F_X(y)$ by means of $P_X$.

**8.51** Use Problem 8.50 and the properties of integrals.

**8.52** Express $P_X((a,b])$ as an integral from $a$ to $b$. Why is the integrand non-negative?

**8.53** You need to verify that the integral from $-\infty$ to $+\infty$ of $f_X$ is 1. Don't try to do it by brute force. Start with $\left(\int_{-\infty}^{+\infty} f_X(x)\, dx\right)^2$, write it as a double integral, change to polar coordinates, and the result will come out quite easily.

**8.54** Express the probability that $a+bX$ belongs to any given Borel set $B$ as an integral over $B$.

**8.55** First, compute the distribution function of $Y$.

**8.56** Look for an increasing function $g$ with values in $[0,1]$.

**8.57** Look at the family of sets $A \subset \mathbb{R}^2$ such that $\{(X,Y) \in A\}$ belongs to $\mathcal{F}$. Is it a $\sigma$-field? Does it contain all rectangles $(a,b) \times (c,d)$? If so, it must contain all Borel sets in $\mathbb{R}^2$, which is what you want to prove.

**8.58** Observe that $\{X \in B\} = \{(X,Y) \in B \times \mathbb{R}\}$.

**8.59** What values can the pair of random variables $(X,Y)$ take? What are the corresponding probabilities? The joint distribution can be represented by listing the probabilities against the possible values of $(X,Y)$. The marginal distributions an either be found directly or by using Problem 8.58.

**8.60** In this case, it may be most convenient to express the joint distribution of $X$ and $Y$ as an array; see Solution 8.59.

**8.61** See Solutions 8.59 and 8.60.

**8.62** If $A$ and $B$ are countable subsets in $\mathbb{R}$, then $A \times B$ is a countable subset in $\mathbb{R}^2$. If $C$ is a countable subset in $\mathbb{R}^2$, then the projections of $C$ onto the coordinate axes are countable subsets in $\mathbb{R}$.

**8.63** The event $\{X = n, Y = m\}$ has probability zero unless $m = 1$ or $n = 1$.

**8.64** See Problem 8.58.

**8.65** Express $P_X((a, b])$ and $P_Y((a, b])$ in terms of the joint density $f_{X,Y}$.

**8.66** Random variables $X$ and $Y$ whose joint distribution $P_{X,Y}$ is concentrated on the line $\Delta = \{(x, x) : x \in \mathbb{R}\}$ in $\mathbb{R}^2$ cannot be jointly continuous (Why?). You can try to look among such random variables.

**8.67** Express the probability that $U \in (a, b]$ and $V \in (c, d]$ as an integral of $f_{X,Y}$ over a suitable set and transform it by a suitable substitution into an integral over $(a, b] \times (c, d]$.

**8.68** First, consider the family of Borel sets $A$ such that the events $\{X \in A\}$ and $\{Y \leqslant b\}$ are independent for every $b \in \mathbb{R}$. Then, consider the family of Borel sets $B$ such that the events $\{X \in A\}$ and $\{Y \in B\}$ are independent for every set $A$ in the first family. Does the first family contain all Borel sets? Does the second family contain all Borel sets?

**8.69** Recall the definition of a Borel function. Consider the events $\{X \in f^{-1}(A)\}$ and $\{Y \in g^{-1}(B)\}$ for any Borel sets $A$ and $B$.

**8.70** Express the probability that $X \leqslant a$ in terms of $f_X$ and the probability that $Y \leqslant b$ in terms of $f_Y$. Then, express the probability that $X \leqslant a$ and $Y \leqslant b$ in terms of the joint density $f_{X,Y}$.

**8.71** a) The integral of the joint density over $\mathbb{R}^2$ is 1.
b) $P(\{X < Y\})$ can be expressed and the integral of the joint density over $\{(x, y) \in \mathbb{R}^2 : x < y\}$.
c) Use (8.3) and (8.4).
d) Use Problem 8.70.

**8.72** Use Van der Monde's formula in Problem 2.7.

**8.73** Use the binomial formula $(\lambda + \mu)^n = \sum_{k=0}^{n} \binom{n}{k} \lambda^k \mu^{n-k}$.

**8.74** Using Problem 8.70, express the distribution function of $X + Y$ in terms of the marginal densities. How can you find the density of $X + Y$ from the distribution function?

## 8.3 Solutions

**8.1** An interval $(a,b)$ is a Borel set, so if $\{X \in B\} \in \mathcal{F}$ for all Borel sets $B$, then $\{X \in (a,b)\} \in \mathcal{F}$ for any interval $(a,b)$.

The converse is based on two observations. First, the family $\mathcal{C}$ of all sets $A \subset \mathbb{R}$ such that $\{X \in A\} \in \mathcal{F}$ is a $\sigma$-field (Problem 5.8). Second, $\mathcal{C}$ contains open intervals, so by the definition of Borel sets (Definition 5.3), $\mathcal{B} \subset \mathcal{C}$, which completes the proof.

**8.2** The only positive value of $X$ is 9, which is produced by the following six outcomes: $(1,1)$, $(1,2)$, $(2,1)$, $(5,6)$, $(6,5)$, $(6,6)$. It follows that

$$P(\{X > 0\}) = P(\{X = 9\}) = \frac{6}{36} = \frac{1}{6},$$

which is the probability of winning.

The only negative value of $X$ is $-10$, which is produced by the following outcomes: $(1,6)$, $(2,5)$, $(3,4)$, $(4,3)$, $(5,2)$, $(6,1)$. As a result, the probability of losing is

$$P(\{X < 0\}) = P(\{X = -10\}) = \frac{6}{36} = \frac{1}{6}.$$

**8.3** When 2 dice are rolled, there are 36 possible outcomes. Out of these, three outcomes $(1,2)$, $(2,1)$, $(2,2)$ will give $X = 2$, five outcomes $(1,3)$, $(2,3)$, $(3,3)$, $(3,2)$, $(3,1)$ will give $X = 3$, and seven outcomes $(1,4)$, $(2,4)$, $(3,4)$, $(4,4)$, $(4,3)$, $(4,2)$, $(4,1)$ will give $X = 4$. Altogether, $\{X \in [2,4]\}$ consists of $3+5+7 = 15$ outcomes, so

$$P(\{X \in [2,4]\}) = \frac{15}{36} = \frac{5}{12}.$$

**8.4** First, we compute the inverse image

$$\{X \in [0, \tfrac{1}{2})\} = \{x \in [0,1] : 0 \leqslant x^2 < \tfrac{1}{2}\} = [0, \tfrac{1}{\sqrt{2}}).$$

But $[0, \frac{1}{\sqrt{2}})$ is an interval, so the Lebesgue measure is equal to its length:

$$P(\{X \in [0, \tfrac{1}{2})\}) = P([0, \tfrac{1}{\sqrt{2}})) = \tfrac{1}{\sqrt{2}}.$$

**8.5** For any positive integer $n$, the probability $P(\{X = n\})$, i.e., the probability that heads will appear for the first time at the $n$th toss of the coin, is $\left(\frac{1}{2}\right)^n$. It follows that

$$\begin{aligned} P(\{X \in 2\mathbb{N}\}) &= \sum_{n=1}^{\infty} P(\{X = 2n\}) = \sum_{n=1}^{\infty} \left(\frac{1}{2}\right)^{2n} \\ &= \sum_{n=1}^{\infty} \left(\frac{1}{4}\right)^{n} = \frac{\frac{1}{4}}{1-\frac{1}{4}} = \frac{1}{3}. \end{aligned}$$

Here, we used the formula for the sum of a geometric progression.

**8.6** Let $B$ be a Borel set. Consider two cases:

1) $c \in B$. Then, $\{X \in B\} = \Omega$.
2) $c \notin B$. Then, $\{X \in B\} = \emptyset$.

But every $\sigma$-field $\mathcal{F}$ must contain $\Omega$ and $\emptyset$, so $X$ must be a random variable with respect to $\mathcal{F}$.

**8.7** Since the inverse image $\{X \in \{3\}\} = \{2\}$ does not belong to $\mathcal{F}$, $X(\omega) = 1 + \omega$ is not a random variable with respect to $\mathcal{F}$.

In the case in hand, for a function $Y$ to be a random variable, it must be constant on the set $\{2, 3, 4\}$ and may take any value at 1. For example, $X(1) = -1$, $X(2) = X(3) = X(4) = 1$ is a random variable.

**8.8** The non-empty inverse images of one-element sets are the atoms that generate the desired smallest $\sigma$-fields.

a) The atoms are $\{0\}$, $\{-1, 1\}$, and $\{-2, 2\}$. They generate the $\sigma$-field

$$\{\emptyset, \Omega, \{0\}, \{-1,1\}, \{-2,2\}, \{-2,-1,1,2\}, \{-2,0,2\}, \{-1,0,1\}\}.$$

b) The atoms are $\{-2\}$, $\{-1\}$, $\{0\}$, $\{1\}$, and $\{2\}$. They generate the $\sigma$-field consisting of all subsets of $\Omega$.

**8.10** We need to decide whether or not the inverse images of Borel sets under $X$ and $Y$ sets belong to $\mathcal{F}$.

a) The inverse image of a Borel set $B \subset \mathbb{R}$ under $X(x) = x$ is

$$\{X \in B\} = \{x \in [0,1] : x \in B\} = B \cap [0,1];$$

that is, it is a Borel set contained in $[0,1]$, which therefore belongs to $\mathcal{F}$. Thus, $X$ is a random variable on $\Omega$ with respect to $\mathcal{F}$.

b) The inverse image of a Borel set $B \subset \mathbb{R}$ under $Y(x) = |x - \frac{1}{2}|$ is

$$\begin{aligned}\{Y \in B\} &= \{x \in [0,1] : |x - \tfrac{1}{2}| \in B\} \\ &= \left((\tfrac{1}{2} - B) \cap [0, \tfrac{1}{2}]\right) \cup \left((\tfrac{1}{2} + B) \cap [\tfrac{1}{2}, 1]\right),\end{aligned}$$

which is also a Borel set contained in $[0,1]$, i.e., it belongs to $\mathcal{F}$. (Here, $\frac{1}{2} + B = \{\frac{1}{2} + x : x \in B\}$ and $\frac{1}{2} - B = \{\frac{1}{2} - x : x \in B\}$.) It follows that $Y$ is a random variable on $\Omega$ with respect to $\mathcal{F}$.

**8.11** In Solution 8.10, the inverse images $\{X \in B\}$ and $\{Y \in B\}$ of any Borel set $B \subset \mathbb{R}$ are computed and shown to be Borel sets contained in $[0,1]$. Do they satisfy the conditions $\{X \in B\} = 1 - \{X \in B\}$ and $\{Y \in B\} = 1 - \{Y \in B\}$?

a) Taking $B = [0, \frac{1}{2}]$, we have

$$\{X \in B\} = B \cap [0,1] = [0, \tfrac{1}{2}],$$

which does not belong to $\mathcal{G}$ since $[0, \frac{1}{2}] \neq 1 - [0, \frac{1}{2}]$. Therefore, $X$ is not a random variable on $\Omega$ with respect to $\mathcal{G}$.

b) For any Borel set $B \subset \mathbb{R}$,

$$\{Y \in B\} = ((\tfrac{1}{2} - B) \cap [0, \tfrac{1}{2}]) \cup ((\tfrac{1}{2} + B) \cap [\tfrac{1}{2}, 1])$$

is a Borel set contained in $[0,1]$ and $\{Y \in B\} = 1 - \{Y \in B\}$, so it belongs to $\mathcal{G}$. It follows that $Y$ is a random variable on $\Omega$ with respect to $\mathcal{G}$.

**8.12** The smallest $\sigma$-field is generated by the inverse images of the values of the random variable. There are $n$ such values, so the field has $n$ atoms. Therefore, it must have $2^n$ elements; see Solution 3.26.

**8.13** For any Borel set $B$, the inverse image $\{X \in B\}$ belongs to $\mathcal{F}$. Since $\mathcal{F} \subset \mathcal{G}$, it follows that $\{X \in B\}$ belongs to $\mathcal{G}$ as well, which proves the claim.

**8.14** Since $\{X \in \mathbb{R}\} = \Omega$, it follows that $P_X(\mathbb{R}) = P(\Omega) = 1$. For any pairwise disjoint Borel sets $A_1, A_2, \ldots$, the inverse images $\{X \in A_1\}, \{X \in A_2\}, \ldots$ are also pairwise disjoint and we have

$$\begin{aligned}
P_X\left(\bigcup_{i=1}^{\infty} A_i\right) &= P\left(\left\{X \in \bigcup_{i=1}^{\infty} A_i\right\}\right) && \text{by the definition of } P_X \\
&= P\left(\bigcup_{i=1}^{\infty}\{X \in A_i\}\right) && \text{by a property of inverse images} \\
&= \sum_{i=1}^{\infty} P(\{X \in A_i\}) && \text{because } P \text{ is countably additive} \\
&= \sum_{i=1}^{\infty} P_X(A_i) && \text{by the definition of } P_X.
\end{aligned}$$

**8.15** Suppose that the distribution of $X$ is a Dirac measure $\delta_a$, where $a \in \mathbb{R}$. This will be so if the function $X : \Omega \to \mathbb{R}$ is constant and equal to $a$. To find a different random variable $X'$ with the same distribution, simply take a different domain. If $\Omega' \neq \Omega$ and $X'$ is identically equal to $a$ on $\Omega'$, then certainly $X$ and $X'$ are different functions.

The above trick sounds cheap, but, in fact, it is not. It emphasizes the relative unimportance of the choice of the underlying space $\Omega$, an essential feature of probability theory. All that really matters is the distribution of a random variable.

**8.16** In Problem 8.2, we computed $P(\{X > 0\})$ and $P(\{X < 0\})$. It follows that

$$P_X((0, +\infty)) = P(\{X > 0\}) = \frac{1}{6}$$

and

$$P_X((-\infty, 0)) = P(\{X < 0\}) = \frac{1}{6}.$$

**8.17** In Problem 8.3, we found $P(\{X \in [2,4]\})$. Thus,

$$P_X([2,4]) = P(\{X \in [2,4]\}) = \frac{5}{12}.$$

**8.18** In Problem 8.4, we found $P(\{X \in [0, \frac{1}{2})\})$. Thus,

$$P_X([0, \tfrac{1}{2})) = P(\{X \in [0, \tfrac{1}{2})\}) = \tfrac{1}{\sqrt{2}}.$$

**8.19** In Problem 8.5, we found $P(\{X \in 2\mathbb{N}\})$. Thus,

$$P_X(2\mathbb{N}) = P(\{X \in 2\mathbb{N}\}) = \frac{1}{3}.$$

**8.20** If $x < c$, then $P(\{X \leqslant x\}) = 0$, since $P(\{X = c\}) = 1$. If $x \geqslant c$, then $P(\{X \leqslant x\}) = 1$. As a result, we obtain the distribution function

$$F_X(x) = \begin{cases} 0 & \text{if } x < c \\ 1 & \text{if } x \geqslant c, \end{cases}$$

whose graph is presented in Figure 8.2.

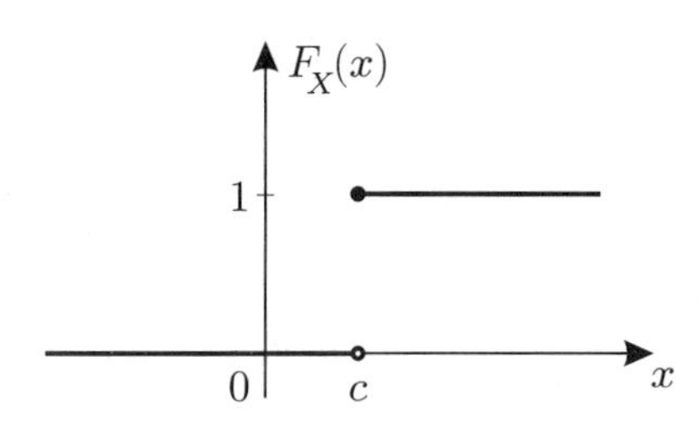


FIGURE 8.2. Distribution function $F_X$ when $X$ is constant, $X(\omega) \equiv c$

**8.21** If $x < 1$, then $P(\{X \leqslant x\}) = 0$. If $1 \leqslant x < 2$, then $P(\{X \leqslant x\}) = \frac{1}{3}$. If $x > 2$, then $P(\{X < x\}) = 1$. The result is the distribution function

$$F_X(x) = \begin{cases} 0 & \text{if } x < 1 \\ \frac{1}{3} & \text{if } 1 \leqslant x < 2 \\ 1 & \text{if } x \geqslant 2, \end{cases}$$

whose graph is presented in Figure 8.3.

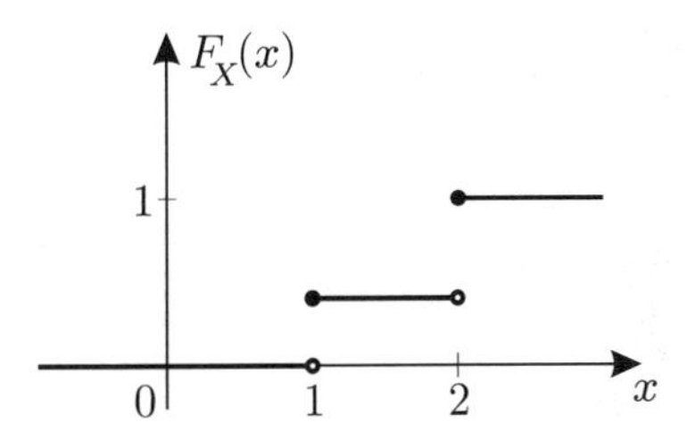


FIGURE 8.3. Distribution function $F_X$ in Problem 8.21

**8.22** If $c_{k-1} \leqslant x < c_k$, then

$$F_X(x) = P(\{X \leqslant x\}) = \bigcup_{j=1}^{k-1} P(\{X = c_j\}) = \sum_{j=1}^{k-1} \alpha_j.$$

If $x < c_1$, then

$$F_X(x) = P(\{X \leqslant x\}) = 0,$$

and if $x \geqslant c_n$, then

$$F_X(x) = P(\{X \leqslant x\}) = 1.$$

**8.24** If $y \geqslant 1$, then $\{X \leqslant y\} = [0, 1]$, so $F_X(y) = P(\{X \leqslant y\}) = 1$. If $y < -1$, then $\{X \leqslant y\} = \emptyset$, so $F_X(y) = P(\{X \leqslant y\}) = 0$.

For $y \in [-1, 1)$, we consider the set $\{x \in [0, 1] : 2x - 1 \leqslant y\}$ to find that it is equal to the interval $[0, \frac{y+1}{2}]$. The Lebesgue measure of this interval is just its length, so $F_X(y) = \frac{y+1}{2}$.

**8.25** The probability $P(\{X = n\}) = \frac{1}{2^n}$ for $n = 1, 2, \ldots$ was found in Solution 8.5. These probabilities add up to 1:

$$\sum_{n=1}^{\infty} P(\{X = n\}) = \sum_{n=1}^{\infty} \frac{1}{2^n} = 1.$$

Thus, the distribution function $F_X$ is given by

$$\begin{aligned}
F_X(x) &= P_X((-\infty, x]) \\
&= P(\{X \in (-\infty, x]\}) \\
&= \sum_{n<x} P(\{X = n\}) \\
&= \begin{cases} 0 & \text{if } x < 1 \\ \frac{1}{2} + \frac{1}{4} + \cdots + \frac{1}{2^n} & \text{if } n \leqslant x < n+1,\ n = 1, 2, \ldots \end{cases} \\
&= \begin{cases} 0 & \text{if } x < 1 \\ 1 - \frac{1}{2^n} & \text{if } n \leqslant x < n+1,\ n = 1, 2, \ldots \end{cases}.
\end{aligned}$$

The graph of $F_X$ is shown in Figure 8.4.

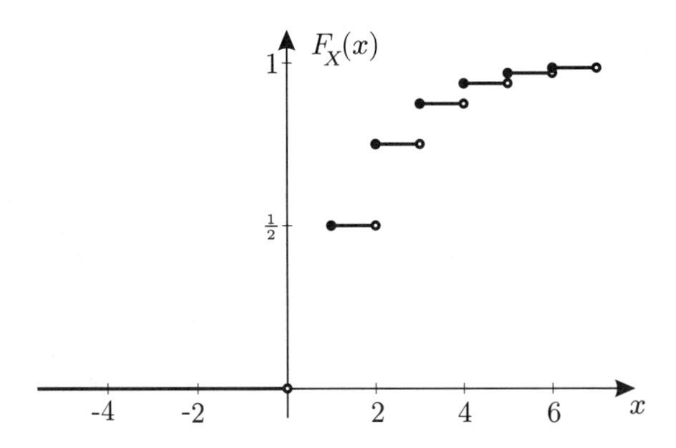


FIGURE 8.4. Distribution function $F_X$ in Problem 8.25

**8.26** If $y_1 \leqslant y_2$, then $\{\omega : X(\omega) \leqslant y_1\} \subset \{\omega : X(\omega) \leqslant y_2\}$. By Problem 4.12,

$$F_X(y_1) = P(\{\omega : X(\omega) \leqslant y_1\}) \leqslant P(\{\omega : X(\omega) \leqslant y_2\}) = F_X(y_2),$$

as required.

**8.27** It is sufficient to show that $F_X(y_n) \to 1$ for any sequence $y_n \nearrow \infty$. Since $F_X(y_n) = P(\{X \leqslant y_n\}) = P(A_n)$, where $A_n = \{X \leqslant y_n\}$, we shall consider the limit of the sequence $P(A_n)$. Since $A_n \subset A_{n+1}$ for each $n$, it follows that $\lim_{n\to\infty} P(A_n) = P(\bigcup_{n=1}^{\infty} A_n)$ by Problem 6.8. But $\bigcup_{n=1}^{\infty} A_n = \Omega$ because for any $\omega \in \Omega$, $X(\omega) \leqslant y_n$ for all sufficiently large $n$. Since $P(\Omega) = 1$, this completes the proof.

**8.28** It is sufficient to show that $F_X(y_n) \to 0$ for any sequence $y_n \searrow -\infty$. Since $F_X(y_n) = P(\{X \leqslant y_n\}) = P(A_n)$, where $A_n = \{X \leqslant y_n\}$, we shall consider the limit of the sequence $P(A_n)$. Since $A_n \supset A_{n+1}$ for each $n$, it follows that $\lim_{n\to\infty} P(A_n) = P(\bigcap_{n=1}^{\infty} A_n)$ by Problem 6.9. But $\bigcap_{n=1}^{\infty} A_n = \emptyset$ because it is impossible to find any $\omega$ such that $X(\omega) \leqslant y_n$ for each $n$. Since $P(\emptyset) = 0$, this completes the proof.

**8.29** If all the values of $X$ are in $[a, b]$ and $y < a$, then $\{X \leqslant y\} = \emptyset$, so $F_X(y) = P(\{X \leqslant y\}) = 0$. If $y \geqslant b$, then $\{X \leqslant y\} = \Omega$ and $F_X(y) = P(\{X \leqslant y\}) = 1$.

**8.30** Fix an arbitrary $y \in \mathbb{R}$ and take any sequence $y_n \searrow y$. The sets $A_n = \{X \leqslant y_n\}$ form a contracting sequence, $A_n \supset A_{n+1}$ for each $n$, so $P(A_n) \to P(\bigcap_{n=1}^{\infty} A_n)$ as $n \to \infty$. It remains to show that $P(\{X \leqslant y\}) = P(\bigcap_{n=1}^{\infty} A_n)$. For this, it is sufficient to show that $\{X \leqslant y\} =$

$\bigcap_{n=1}^{\infty} A_n$. If $X(\omega) \leqslant y$ for some $\omega$, then $X(\omega) \leqslant y < y_n$ for each $n$; hence, $\omega \in \bigcap_{n=1}^{\infty} A_n$. On the other hand, if $\omega \in \bigcap_{n=1}^{\infty} A_n$, then $\omega \in A_n$ for each $n$. This means that $X(\omega) \leqslant y_n$ and so $X(\omega) \leqslant y$, which completes the proof.

**8.31** By Problem 8.30, we only need to show that $F_X$ is left-continuous at $y$ if and only if $P_X(\{y\}) = 0$.

Fix $y$ and take any sequence $y_n \nearrow y$. Then, $\{X \leqslant y_n\}$ is an expanding sequence of events with union $\{X < y\}$. It follows that

$$F_X(y_n) = P(\{X \leqslant y_n\}) \to P(\{X < y\}) = F_X(y) - P_X(\{y\}).$$

Therefore, $F_X$ is left-continuous at $y$ if and only if $P_X(\{y\}) = 0$.

**8.32** a) We have $(a, b] = (-\infty, b] \setminus (-\infty, a]$ and $(-\infty, a] \subset (-\infty, b]$. Since $P_X$ is a measure (Problem 8.14),

$$P_X((-\infty, b] \setminus (-\infty, a]) = P_X((-\infty, b]) - P_X((-\infty, a])$$

by Problem 4.10. Hence,

$$P_X((a, b]) = F_X(b) - F_X(a).$$

b) We have $(a, b) = \bigcup_{n=1}^{\infty} (a, b - \frac{1}{n}]$ and

$$\begin{aligned} P_X((a, b)) &= P_X\left(\textstyle\bigcup_{n=1}^{\infty} (a, b - \frac{1}{n}]\right) \\ &= \lim_{n\to\infty} P_X((a, b - \tfrac{1}{n}]) && \text{by Problem 6.8} \\ &= \lim_{n\to\infty} \left(F_X(b - \tfrac{1}{n}) - F_X(a)\right) && \text{by a)} \\ &= \lim_{y \nearrow b} F_X(y) - F_X(a), && \text{since } F_X \text{ is monotone.} \end{aligned}$$

c) We have $[a, b] = \bigcap_{n=1}^{\infty} (a - \frac{1}{n}, b]$ and

$$\begin{aligned} P_X([a, b]) &= P_X\left(\textstyle\bigcap_{n=1}^{\infty} (a - \frac{1}{n}, b]\right) \\ &= \lim_{n\to\infty} P_X((a - \tfrac{1}{n}, b]) && \text{by Problem 6.9} \\ &= \lim_{n\to\infty} \left(F_X(b) - F_X(a - \tfrac{1}{n})\right) && \text{by a)} \\ &= F_X(b) - \lim_{y \nearrow a} F_X(y), && \text{since } F_X \text{ is monotone.} \end{aligned}$$

d) Since $[a, b) = \bigcup_{n=1}^{\infty} [a, b - \frac{1}{n}]$,

$$\begin{aligned} P_X([a, b)) &= P_X\left(\textstyle\bigcup_{n=1}^{\infty} [a, b - \frac{1}{n}]\right) \\ &= \lim_{n\to\infty} P_X\left([a, b - \tfrac{1}{n}]\right) && \text{by Problem 6.8} \\ &= \lim_{n\to\infty} \left(F_X(b - \tfrac{1}{n}) - \lim_{y \nearrow a} F_X(y)\right) && \text{by c)} \\ &= \lim_{y \nearrow b} F_X(y) - \lim_{y \nearrow a} F_X(y), && \text{since } F_X \text{ is monotone.} \end{aligned}$$

**8.33** For $x < 0$, $P(\{X \leqslant x\}) = 0$ because distance cannot be negative. For $x \geqslant \frac{1}{2}$, $P(X \leqslant x) = 1$ because the greatest value of $X$ is $\frac{1}{2}$ (when we select the center of the square). For $x \in [0, \frac{1}{2})$, it is easy to find that $P(\{X > x\}) = (1 - 2x)^2$; see Figure 8.1. It follows that $F_X(x) = 1 - (1 - 2x)^2$ for $x \in [0, \frac{1}{2})$. The graph of $F_X$ is shown in Figure 8.5.

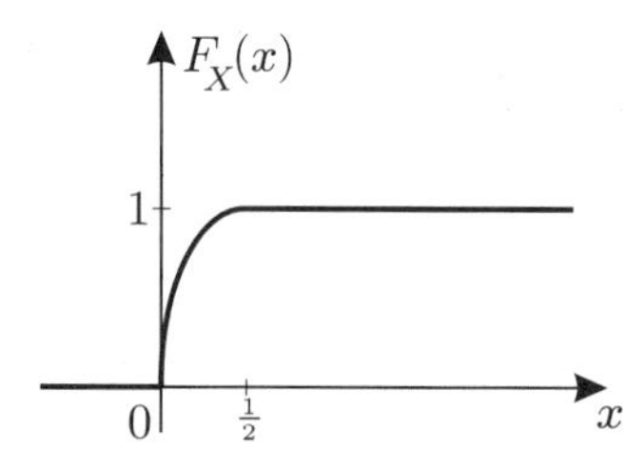


FIGURE 8.5. Graph of $F_X$ in Problem 8.33

**8.34** For $x < -2$,

$$F_X(x) = P(\{X \leqslant x\}) = 0.$$

For $-2 \leqslant x < 0$,

$$F_X(x) = P(\{X \leqslant x\}) = P(\{X = -2\}) = \frac{1}{2}.$$

For $0 \leqslant x < 10$,

$$P(\{X \leqslant x\}) = \frac{1}{2} + \frac{1}{2} \times \frac{x}{10},$$

since either $X = -2$ or the wheel shows a number between 0 and $x$ with probability $\frac{x}{10}$, which has to be multiplied by $\frac{1}{2}$, the probability that the wheel will be used at all. Finally, for $x \geqslant 10$,

$$F_X(x) = P(\{X \leqslant x\}) = 1.$$

The graph of $F_X$ is shown in Figure 8.6.

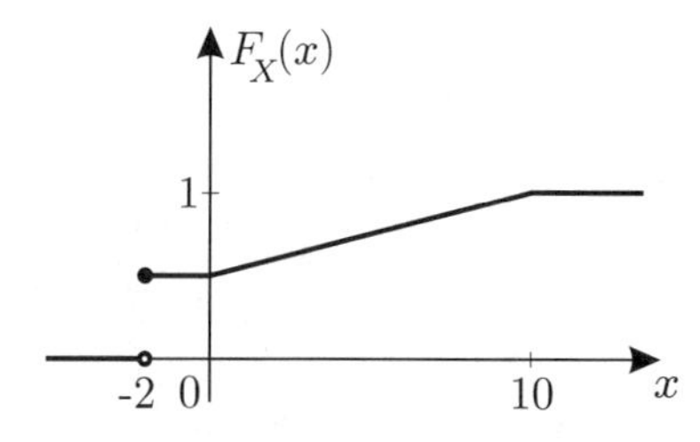


FIGURE 8.6. Graph of $F_X$ in Problem 8.34

Now, we can compute

$$\begin{aligned}P(\{X \geqslant 5\}) &= 1 - P(\{X < 5\}) \\ &= 1 - \lim_{x \nearrow 5} F_X(x) = 1 - F_X(5) = 1 - \frac{3}{4} = \frac{1}{4}.\end{aligned}$$

**8.35** With probability $\frac{1}{2}$, you will not repair your car and then $X = 100$. If you fix the car at time $x$ between 12 and 1 o'clock, the rescue car will have driven $100x$ miles by then. If you start traveling at the same speed as the team, you will meet them in the middle of the remaining distance, so $X = \frac{1}{2}(100 + 100x) = 50x + 50$. The probability that $X < 50x + 50$ is $\frac{1}{2}x$. Therefore,

$$F_X(y) = P(\{X \leqslant y\}) = \begin{cases} 0 & \text{if } y < 50 \\ \frac{y-50}{100} & \text{if } 50 \leqslant y < 100 \\ 1 & \text{if } y \geqslant 100. \end{cases}$$

Note the discontinuity at $y = 100$.

**8.36** The sample space is $\Omega = [0,1] \times [0,1]$ with uniform probability $P$ (which measures the surface area). Here, $(t_1, t_2) \in \Omega$ are the departure times of car 1 and car 2, respectively. Consider the following events:

1) $\{t_2 - t_1 \leqslant -\frac{1}{2}\} \subset \Omega$.
   In this case, the cars will meet at A, so $X = 0$.
2) $\{-\frac{1}{2} < t_2 - t_1 \leqslant 0\} \subset \Omega$.
   This means that car 2 leaves first and will have driven for $t_1 - t_2$ hours, covering $100(t_1 - t_2)$ miles before car 1 starts from A. They will meet in the middle of the remaining distance, i.e., $X = \frac{1}{2}[50 - 100(t_1 - t_2)] = 25 + 50(t_2 - t_1)$ miles away from town A.
3) $\{0 < t_2 - t_1 \leqslant \frac{1}{2}\} \subset \Omega$.
   This means that car 1 leaves first and will have driven for $t_2 - t_1$ hours, covering $100(t_2 - t_1)$ miles before car 2 starts from B. They will meet in the middle of the remaining distance, i.e., $X = 100(t_2 - t_1) + \frac{1}{2}[50 - 100(t_2 - t_1)] = 25 + 50(t_2 - t_1)$ miles away from town A.
4) $\{\frac{1}{2} < t_2 - t_1\} \subset \Omega$.
   In this case, the cars will meet at B, so $X = 50$.

Now, we can compute the distribution function. From the above, we have

$$\{X \leqslant x\} = \begin{cases} \emptyset & \text{if } x < 0 \\ \{25 + 50(t_2 - t_1) \leqslant x\} & \text{if } 0 \leqslant x < 50 \\ \Omega & \text{if } 50 \leqslant x. \end{cases}$$

Examples of such sets are shown in Figure 8.7.

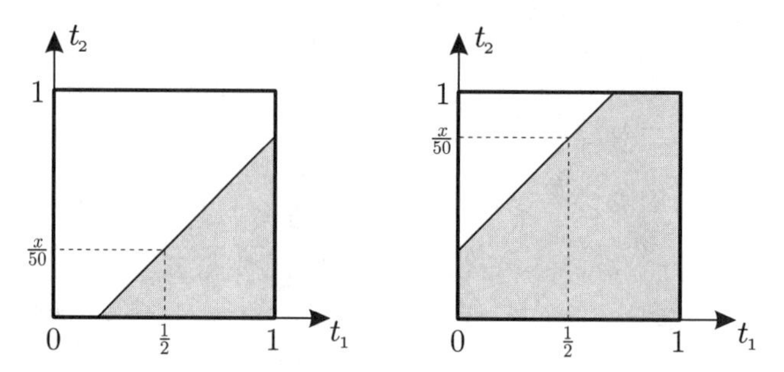


FIGURE 8.7. The sets $\{X < x\}$ for $0 < x < 25$ and $25 < x < 50$

It follows that

$$F_X(x) = P(\{X \leqslant x\}) = \begin{cases} 0 & \text{if } x < 0 \\ \frac{1}{2}\left(\frac{1}{2} + \frac{x}{50}\right)^2 & \text{if } 0 \leqslant x < 25 \\ 1 - \frac{1}{2}\left(\frac{3}{2} - \frac{x}{50}\right)^2 & \text{if } 25 \leqslant x < 50 \\ 1 & \text{if } 50 \leqslant x; \end{cases}$$

see Figure 8.8.

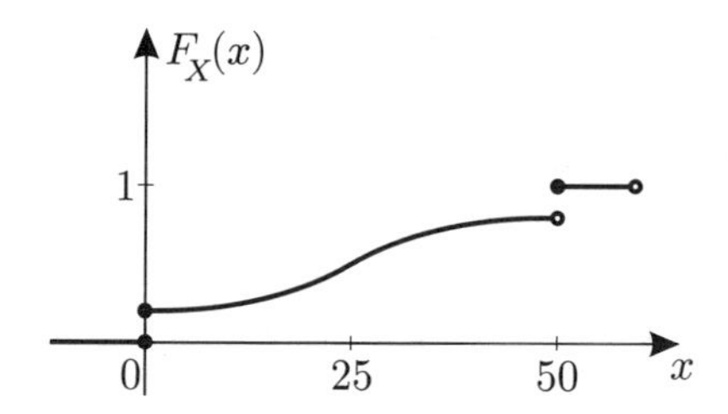


FIGURE 8.8. Graph of $F_X$ in Problem 8.36

**8.37** Suppose that $X$ is a discrete random variable. Then, there is a countable set $C = \{x_1, x_2, \dots\} \subset \mathbb{R}$ such that $P_X(C) = 1$. We can assume without loss of generality that $x_j \neq x_k$ if $j \neq k$. Since $P_X(C) = 1$,

$$\begin{aligned} P_X(B) = P_X(B \cap C) &= \sum_{k=1}^{\infty} P_X(B \cap \{x_k\}) \\ &= \sum_{k=1}^{\infty} P_X(\{x_k\})\delta_{x_k}(B) = \sum_{k=1}^{\infty} \alpha_k \delta_{x_k}(B), \end{aligned}$$

for any Borel set $B$ in $\mathbb{R}$, where $\alpha_k = P_X(\{x_k\}) \geqslant 0$ for $k = 1, 2, \dots$ and where $\delta_{x_k}$ is the Dirac measure concentrated at $x_k$; see Definition 6.6. Moreover,

$$\sum_{k=1}^{\infty} \alpha_k = \sum_{k=1}^{\infty} P_X(\{x_k\}) = P_X(C) = 1,$$

which proves that $P_X$ is a discrete probability measure on $\mathbb{R}$ (see Definition 6.7).

Conversely, suppose that $P_X$ is a discrete probability measure on $\mathbb{R}$; that is,

$$P_X = \sum_{k=1}^{\infty} \alpha_k \delta_{x_k}$$

for some $\alpha_k \geqslant 0$ and $x_k \in \mathbb{R}$, where $k = 1, 2, \ldots$, such that $\sum_{k=1}^{\infty} \alpha_k = 1$. We put $C = \{x_1, x_2, \ldots\}$, which is clearly a countable subset in $\mathbb{R}$. Moreover,

$$P_X(C) = \sum_{k=1}^{\infty} \alpha_k \delta_{x_k}(C) = \sum_{k=1}^{\infty} \alpha_k = 1,$$

so $X$ is a discrete random variable.

**8.38** Let $X$ be a random variable with binomial distribution $B(n, p)$. We put $C = \{0, 1, 2, \ldots, n\}$, which is clearly a countable set (in fact, a finite set). Then,

$$P_X(C) = \sum_{k=0}^{n} P_X(\{k\}) = \sum_{k=0}^{n} \binom{n}{k} p^k (1-p)^{n-k} = (p + 1 - p)^n = 1,$$

which means that $X$ is a discrete random variable.

**8.39** Clearly,

$$P_{\mathbb{I}_A}(\{1\}) = P(\{\mathbb{I}_A = 1\}) = P(A) = p,$$
$$P_{\mathbb{I}_A}(\{0\}) = P(\{\mathbb{I}_A = 0\}) = P(\Omega \setminus A) = 1 - p,$$

where $p = P(A)$. Therefore, $\mathbb{I}_A$ has the Bernoulli distribution $B(1, p)$.

**8.40** By (8.1),

$$\begin{aligned} P_X(\{k\}) &= \binom{n}{k} p^k (1-p)^{n-k} \\ &= \frac{p}{1-p} \frac{n-k+1}{k} \binom{n}{k-1} p^{k-1} (1-p)^{n-k+1} \\ &= \frac{p}{1-p} \frac{n-k+1}{k} P_X(\{k-1\}). \end{aligned}$$

**8.41** By (8.1),

$$\begin{aligned} P_X(\{k\}) &= \binom{n}{k} p^k (1-p)^{n-k} \\ &= \binom{n}{n-k} (1-p)^{n-k} p^k = P_Y(\{n-k\}). \end{aligned}$$

**8.42** There are $2^n$ possible outcomes, each represented by an $n$-element sequence consisting of heads and tails. Among those there are $\binom{n}{k}$ outcomes with $k$ heads and $n-k$ tails; see Problem 2.3. The probability of any given outcome with $k$ heads and $n-k$ tails is $p^k(1-p)^{n-k}$. It follows that

$$P(\{X = k\}) = \binom{n}{k} p^k (1-p)^{n-k},$$

so $X$ has the binomial distribution $B(n, p)$.

**8.43** a) The event $\{Y = n\}$, i.e., "the number of flips is $n$" is the intersection of two independent events: "the first $n-1$ flips include $k-1$ heads" and "the $n$th flip is a head." The probabilities of these events are $\binom{n-1}{k-1} p^{k-1}(1-p)^{n-k}$ (see Problem 8.42) and $p$, respectively. By multiplying these probabilities, we find

$$P_Y(\{n\}) = P(\{Y = n\}) = \binom{n-1}{k-1} p^k (1-p)^{n-k}.$$

b) Clearly,

$$P_Y = \sum_{n=k}^{\infty} \alpha_n \delta_{x_n},$$

where $x_n = n$ and $\alpha_n = \binom{n-1}{k-1} p^k (1-p)^{n-k} \geqslant 0$ for $n = k, k+1, k+2, \dots$. We only need to verify that the $\alpha_n$ add up to 1:

$$\begin{aligned}
\sum_{n=k}^{\infty} \alpha_n &= \sum_{n=k}^{\infty} \binom{n-1}{k-1} p^k (1-p)^{n-k} \\
&= p^k \sum_{n=k}^{\infty} \binom{n-1}{k-1} (1-p)^{n-k} = p^k \frac{1}{p^k} = 1.
\end{aligned}$$

To compute the sum, we put $x = 1 - p$ in the Taylor expansion $\sum_{n=k}^{\infty} \binom{n-1}{k-1} x^{n-k}$ of $(1-x)^{-k}$ around 0.

**8.44** By (8.1),

$$\begin{aligned}
P(\{X_n = k\}) &= \binom{n}{k} \left(\frac{\lambda}{n}\right)^k \left(1 - \frac{\lambda}{n}\right)^{n-k} \\
&= \frac{n!}{(n-k)! n^k} \left(1 - \frac{\lambda}{n}\right)^{-k} \left(1 - \frac{\lambda}{n}\right)^n \frac{\lambda^k}{k!} \to e^{-\lambda} \frac{\lambda^k}{k!},
\end{aligned}$$

since $\frac{n!}{(n-k)!n^k} \to 1$, $\left(1 - \frac{\lambda}{n}\right)^{-k} \to 1$, and $\left(1 - \frac{\lambda}{n}\right)^n \to e^{-\lambda}$ as $n \to \infty$.

**8.45** Suppose that $X$ has the Poisson distribution with parameter $\lambda$. Put $C = \{0, 1, 2, \dots\}$, which is clearly a countable set. Then,

$$P_X(C) = \sum_{k=0}^{\infty} P_X(\{k\}) = \sum_{k=0}^{\infty} e^{-\lambda}\frac{\lambda^k}{k!} = e^{-\lambda}\sum_{k=0}^{\infty}\frac{\lambda^k}{k!} = e^{-\lambda}e^{\lambda} = 1,$$

since $\sum_{k=0}^{\infty}\frac{\lambda^k}{k!}$ is the Taylor expansion of $e^{\lambda}$ around 0. This means that $X$ is a discrete random variable.

**8.46** Let $X$ be the number of particles reaching the safety device within a 1-second period. The alarm will be activated if $X \geqslant 5$. The probability of this is

$$\begin{aligned} P(\{X \geqslant 5\}) &= 1 - P(\{X < 5\}) \\ &= 1 - e^{-\lambda}\left(1 + \frac{\lambda}{1!} + \frac{\lambda^2}{2!} + \frac{\lambda^3}{3!} + \frac{\lambda^4}{4!}\right) \approx 0.00017 \end{aligned}$$

if $\lambda = 0.5$ .

**8.47** By (8.2),

$$P_X(\{k+1\}) = e^{-\lambda}\frac{\lambda^{k+1}}{(k+1)!} = \frac{\lambda}{k+1}e^{-\lambda}\frac{\lambda^k}{k!} = \frac{\lambda}{k+1}P_X(\{k\}).$$

**8.48** By definition, $\int_{-n}^{n} f_X(x)\,\mathrm{d}x = P_X((-n, n])$. The intervals $(-n, n]$ form an expanding sequence of sets with union $\mathbb{R}$. By Problem 6.8,

$$P_X(\mathbb{R}) = \lim_{n\to\infty} P_X((-n, n]),$$

but $P_X(\mathbb{R}) = 1$, so

$$\begin{aligned} \int_{-\infty}^{\infty} f_X(x)\,\mathrm{d}x &= \lim_{n\to\infty}\int_{-n}^{n} f_X(x)\,\mathrm{d}x \\ &= \lim_{n\to\infty} P_X(A_n) = 1. \end{aligned}$$

**8.49** We have $\{y\} = \bigcap_n (y - \frac{1}{n}, y]$ and the result follows from Problem 6.9. Bear in mind that

$$P_X((y - \tfrac{1}{n}, y]) = \int_{y-\frac{1}{n}}^{y} f_X(x)\,\mathrm{d}x \to \int_{y}^{y} f_X(x)\,\mathrm{d}x = 0$$

as $n \to \infty$, which holds because the function $z \mapsto \int_z^y f_X(x)\,\mathrm{d}x$ is continuous.

**8.50** By the definition of $F_X$ and Problem 6.8,

$$\begin{aligned} F_X(y) = P_X((-\infty, y]) &= \lim_{n\to\infty} P_X((-n, y]) \\ &= \lim_{n\to\infty}\int_{-n}^{y} f_X(x)\,\mathrm{d}x = \int_{-\infty}^{y} f_X(x)\,\mathrm{d}x. \end{aligned}$$

This is because the intervals $(-n, y]$ form an expanding sequence of sets whose union is $(-\infty, y]$.

**8.51** Since, by Problem 8.50, $F_X(y) = \int_{-\infty}^{y} f_X(x)\,dx$ and the function $y \mapsto \int_{-\infty}^{y} f_X(x)\,dx$ is continuous, the claim follows.

Alternatively, Problems 8.31 and 8.49 can be applied to obtain the same result.

**8.52** Since $F_X$ is non-decreasing, $f_X = \frac{dF_X}{dx}$ is non-negative. By Problem 8.32 a),

$$P_X((a,b]) = F_X(a) - F_X(b) = \int_a^b \frac{dF_X(x)}{dx}\,dx = \int_a^b f_X(x)\,dx.$$

This means that $X$ is an absolutely continuous random variable with density $f_X$.

**8.53** Clearly, $f_X(x) \geqslant 0$ for all $x \in \mathbb{R}$. We need to verify that the integral from $-\infty$ to $+\infty$ of $f_X$ is 1. Let us begin with $m = 0$ and $\sigma = 1$. Then,

$$\begin{aligned}
&\left(\int_{-\infty}^{+\infty} \frac{1}{\sqrt{2\pi}} \exp\left(-\frac{x^2}{2}\right) dx\right)^2 \\
&\qquad = \frac{1}{2\pi} \int_{-\infty}^{+\infty} \int_{-\infty}^{+\infty} \exp\left(-\frac{x^2+y^2}{2}\right) dx\,dy \\
&\qquad = \frac{1}{2\pi} \int_0^{\infty} \int_0^{2\pi} r \exp\left(-\frac{r^2}{2}\right) d\varphi\,dr \\
&\qquad = \int_0^{\infty} r \exp\left(-\frac{r^2}{2}\right) dr \\
&\qquad = -\exp\left(-\frac{r^2}{2}\right)\Big|_0^{\infty} = 1.
\end{aligned}$$

To compute the above double integral, we substituted the polar coordinates $r$ and $\varphi$ for $x$ and $y$.

Now, for any $a$ and $b$,

$$\begin{aligned}
&\int_{-\infty}^{+\infty} \frac{1}{\sqrt{2\pi\sigma^2}} \exp\left(-\frac{(x-m)^2}{2\sigma^2}\right) dx \\
&\qquad = \int_{-\infty}^{+\infty} \frac{1}{\sqrt{2\pi}} \exp\left(-\frac{z^2}{2}\right) dz = 1,
\end{aligned}$$

where the substitution $z = m + \sigma x$ is used. It follows that

$$\int_{-\infty}^{+\infty} f_X(x)\,dx = 1,$$

as required.

**8.54** For any Borel set $B$ in $\mathbb{R}$,

$$\begin{aligned} P(\{a+bX \in B\}) &= P(\{X \in (B-a)/b\}) \\ &= \int_{(B-a)/b} \frac{1}{\sqrt{2\pi}} \exp\left(-\frac{x^2}{2}\right) dx \\ &= \int_B \frac{1}{\sqrt{2\pi b^2}} \exp\left(-\frac{(y-a)^2}{2b^2}\right) dx, \end{aligned}$$

which means that $a+bX$ has the normal distribution $N(a, b^2)$. The integral above has been transformed by substituting $y = a + bx$.

**8.55** The distribution function of $Y$ is

$$\begin{aligned} F_Y(y) &= P(\{Y \leqslant y\}) = P(\{e^X \leqslant y\}) \\ &= P(\{X \leqslant \ln y\}) = \int_0^{\ln y} \frac{1}{\sqrt{2\pi}} \exp\left(-\frac{x^2}{2}\right) dx. \end{aligned}$$

It follows that

$$f_Y(y) = \frac{dF_Y(y)}{dy} = \frac{1}{y} \frac{1}{\sqrt{2\pi}} \exp\left(-\frac{\ln^2 y}{2}\right) = \frac{1}{\sqrt{2\pi y^2 y^{\ln y}}}.$$

**8.56** We shall be looking for an increasing function $g : \mathbb{R} \to [0,1]$. For such a function,

$$P(\{g(X) \leqslant y\}) = P(\{X \leqslant g^{-1}(y)\}) = \int_0^{g^{-1}(y)} f_X(x)\, dx = g^{-1}(y),$$

since $f_X(x) = 1$ for any $x \in [0,1]$. If $g(Y)$ is to have the normal distribution $N(0,1)$, then

$$g^{-1}(y) = P(\{g(X) \leqslant y\}) = \int_{-\infty}^{y} \frac{1}{\sqrt{2\pi}} \exp\left(-\frac{x^2}{2}\right) dx.$$

This means that $g$ is the inverse to the distribution function of the normal distribution $N(0,1)$.

**8.57** Denote by $\mathcal{G}$ the family of sets $A \subset \mathbb{R}^2$ such that $\{(X,Y) \in A\}$ belongs to $\mathcal{F}$. We need to show that all Borel sets in $\mathbb{R}^2$ belong to $\mathcal{G}$.

Clearly, every rectangle $(a,b) \times (c,d)$ is in $\mathcal{G}$, since both $\{X \in (a,b)\}$ and $\{Y \in (c,d)\}$ belong to $\mathcal{F}$ and so

$$\{(X,Y) \in (a,b) \times (c,d)\} = \{X \in (a,b)\} \cap \{Y \in (c,d)\}$$

also belongs to $\mathcal{F}$. Moreover, $\mathcal{G}$ is clearly a $\sigma$-field. Because the $\sigma$-field of Borel sets in $\mathbb{R}^2$ is generated by rectangles, it follows that it is contained in $\mathcal{G}$, as required.

**8.58** For any Borel set $B$ in $\mathbb{R}$,

$$\begin{aligned} P_X(B) &= P(\{X \in B\}) = P(\{X \in B\} \cap \{Y \in \mathbb{R}\}) \\ &= P(\{(X, Y) \in B \times \mathbb{R}\}) = P_{X,Y}(B \times \mathbb{R}), \end{aligned}$$

since $\{Y \in \mathbb{R}\} = \Omega$. The formula for $P_Y(B)$ can be obtained in a similar manner.

**8.59** The pair $(X, Y)$ can take the values $(0, 2)$, $(1, 1)$, and $(2, 0)$ with probabilities $\frac{1}{2}$, $\frac{1}{4}$, and $\frac{1}{2}$, respectively. The joint distribution of $X$ and $Y$ can be expressed as

$$P_{X,Y} = \frac{1}{4}\delta_{0,2} + \frac{1}{2}\delta_{1,1} + \frac{1}{4}\delta_{2,0}\,,$$

where $\delta_{x,y}$ denotes the Dirac measure on $\mathbb{R}^2$ concentrated at a point $(x, y)$; that is, $\delta_{x,y}(B) = 1$ if $(x, y) \in B$ and $\delta_{x,y}(B) = 0$ if $(x, y) \in B$, for any Borel set $B$ in $\mathbb{R}^2$. The marginal distributions are

$$\begin{aligned} P_X &= \frac{1}{4}\delta_0 + \frac{1}{2}\delta_1 + \frac{1}{4}\delta_2\,, \\ P_Y &= \frac{1}{4}\delta_2 + \frac{1}{2}\delta_1 + \frac{1}{4}\delta_0\,. \end{aligned}$$

Another convenient way of representing the joint distribution of $X$ and $Y$ is in the form of an array:

| | | $X$: 0 | 1 | 2 | |
|---|---|---|---|---|---|
| $Y$ | 0 | 0 | 0 | $\frac{1}{4}$ | $\frac{1}{4}$ |
| | 1 | 0 | $\frac{1}{2}$ | 0 | $\frac{1}{2}$ |
| | 2 | $\frac{1}{4}$ | 0 | 0 | $\frac{1}{4}$ |
| | | $\frac{1}{4}$ | $\frac{1}{2}$ | $\frac{1}{4}$ | |

Here, the columns and rows are labeled by the values of $X$ and $Y$, respectively. The main body of the array represents the joint distribution The column and row totals give the marginal distributions. (The fact that the totals are usually placed in the right and bottom margins of the array explains the term "marginal distributions.")

**8.60** The joint distribution of $X$ and $Y$ can be expressed as an array:

| | | $X$: 1 | 2 | 3 | 4 | 5 | 6 | |
|---|---|---|---|---|---|---|---|---|
| $Y$ | 1 | $\frac{1}{36}$ | $\frac{2}{36}$ | $\frac{2}{36}$ | $\frac{2}{36}$ | $\frac{2}{36}$ | $\frac{2}{36}$ | $\frac{11}{36}$ |
| | 2 | 0 | $\frac{1}{36}$ | $\frac{2}{36}$ | $\frac{2}{36}$ | $\frac{2}{36}$ | $\frac{2}{36}$ | $\frac{9}{36}$ |
| | 3 | 0 | 0 | $\frac{1}{36}$ | $\frac{2}{36}$ | $\frac{2}{36}$ | $\frac{2}{36}$ | $\frac{7}{36}$ |
| | 4 | 0 | 0 | 0 | $\frac{1}{36}$ | $\frac{2}{36}$ | $\frac{2}{36}$ | $\frac{5}{36}$ |
| | 5 | 0 | 0 | 0 | 0 | $\frac{1}{36}$ | $\frac{2}{36}$ | $\frac{3}{36}$ |
| | 6 | 0 | 0 | 0 | 0 | 0 | $\frac{1}{36}$ | $\frac{1}{36}$ |
| | | $\frac{1}{36}$ | $\frac{3}{36}$ | $\frac{5}{36}$ | $\frac{7}{36}$ | $\frac{9}{36}$ | $\frac{11}{36}$ | |

The column and row totals give the marginal distributions.

**8.61** The joint distribution of $X$ and $Y$ can be expressed as an array:

| | | $X$ | | | |
|---|---|---|---|---|---|
| | | 0 | 1 | 2 | |
| $Y$ | 0 | 0 | $\frac{2}{15}$ | $\frac{1}{15}$ | $\frac{1}{5}$ |
| | 1 | $\frac{1}{5}$ | $\frac{2}{5}$ | 0 | $\frac{3}{5}$ |
| | 2 | $\frac{1}{5}$ | 0 | 0 | $\frac{1}{5}$ |
| | | $\frac{2}{5}$ | $\frac{8}{15}$ | $\frac{1}{15}$ | |

The column and row totals give the marginal distributions.

**8.62** Suppose that $X$ and $Y$ are jointly discrete random variables. Then, there is a countable subset $C = \{(x_1, y_1), (x_2, y_2), \dots\}$ in $\mathbb{R}^2$ such that $P_{X,Y}(C) = 1$. Put $A = \{x_1, x_2, \dots\}$ and $B = \{y_1, y_2, \dots\}$. Obviously, these are countable subsets in $\mathbb{R}$. We claim that $P_X(A) = P_Y(B) = 1$, which means that both $X$ and $Y$ are discrete random variables. To verify the claim, observe that $A \times \mathbb{R} \supset C$, so

$$1 \geqslant P_X(A) = P_{X,Y}(A \times \mathbb{R}) \geqslant P_{X,Y}(C) = 1$$

by Problems 8.37 and 4.12. Similarly, since $\mathbb{R} \times B \supset C$, we can prove that $P_Y(B) = 1$.

Now, suppose that $X$ and $Y$ are discrete random variables. This means that there are countable subsets $A$ and $B$ in $\mathbb{R}$ such that $P_X(A) = P_Y(B) = 1$. We take $C = A \times B$, which is a countable subset in $\mathbb{R}^2$. Then, $P_{X,Y}(A \times \mathbb{R}) = P_X(A) = 1$ and $P_{X,Y}(\mathbb{R} \times B) = P_Y(B) = 1$ by Problem 8.58. Now, by Problem 4.27,

$$P_{X,Y}(C) = P_{X,Y}(A \times B) = P_{X,Y}((A \times \mathbb{R}) \cap (\mathbb{R} \times B)) = 1.$$

It follows that $X$ and $Y$ are jointly discrete.

**8.63** We shall compute the values of $P_{X,Y}$ at points of the form $(1, n)$ or $(n, 1)$, where $n = 2, 3, \dots$; namely

$$\begin{aligned} P_{X,Y}(\{(1,n)\}) &= P(\{X = 1, Y = n\}) \\ &= P(\{Y = n\}) = \frac{1}{2^n}, \qquad (8.6) \\ P_{X,Y}(\{(n,1)\}) &= P(\{X = n, Y = 1\}) \\ &= P(\{Y = n\}) = \frac{1}{2^n}, \qquad (8.7) \end{aligned}$$

since $\{X = 1\} \subset \{Y = n\}$ and $\{Y = 1\} \subset \{X = n\}$ for any $n = 2, 3, \dots$ .

Taking $C = \{(1,n) : n = 2,3,\dots\} \cup \{(n,1) : n = 2,3,\dots\}$, which is obviously a countable set, we have

$$\begin{aligned} P_{X,Y}(C) &= \sum_{n=2}^{\infty} P_{X,Y}(\{(1,n)\}) + \sum_{n=2}^{\infty} P_{X,Y}(\{(n,1)\}) \\ &= \sum_{n=2}^{\infty} \frac{1}{2^n} + \sum_{n=2}^{\infty} \frac{1}{2^n} = \frac{1}{2} + \frac{1}{2} = 1. \end{aligned}$$

This means that the joint distribution of $X$ and $Y$ is discrete and determined by (8.6) and (8.7).

The marginal distributions are given by

$$P_X(\{n\}) = P_Y(\{n\}) = \frac{1}{2^n}$$

for $n = 1,2,\dots$ .

**8.64** The marginal distribution $P_X$ can be found as follows:

$$\begin{aligned} P_X(\{m\}) &= P_{X,Y}(\{m\} \times \mathbb{R}) \\ &= \sum_{n=1}^{\infty} P_{X,Y}(\{(m,n)\}) = \sum_{n=1}^{m} \frac{1}{2^{m+1}} = \frac{m}{2^{m+1}} \end{aligned}$$

for any $m = 1,2,\dots$ . The marginal distribution $P_Y$ is

$$\begin{aligned} P_Y(\{n\}) &= P_{X,Y}(\mathbb{R} \times \{n\}) \\ &= \sum_{m=1}^{\infty} P_{X,Y}(\{(m,n)\}) = \sum_{m=n}^{\infty} \frac{1}{2^{m+1}} = \frac{1}{2^n} \end{aligned}$$

for any $n = 1,2,\dots$ .

**8.65** If $X$ and $Y$ are jointly continuous random variables with joint density $f_{X,Y}$, then for any $a < b$,

$$P_X((a,b]) = P_{X,Y}((a,b] \times \mathbb{R}) = \int_a^b \left( \int_{-\infty}^{+\infty} f_{X,Y}(x,y)\, \mathrm{d}y \right) \mathrm{d}x,$$

$$P_Y((a,b]) = P_{X,Y}(\mathbb{R} \times (a,b]) = \int_a^b \left( \int_{-\infty}^{+\infty} f_{X,Y}(x,y)\, \mathrm{d}x \right) \mathrm{d}y,$$

so $X$ and $Y$ are absolutely continuous random variables with density $f_X$ and $f_Y$ as in (8.3) and (8.4).

**8.66** Take $X$ to be an absolutely continuous random variable with density $f_X(x) = 1$ for each $x \in [0,1]$ and 0 otherwise, and put $X = Y$.

To see that $X$ and $Y$ are not jointly continuous, consider the line $\Delta = \{(x,x) : x \in \mathbb{R}\}$ in $\mathbb{R}^2$. Then,

$$P(\{(X,Y) \notin \Delta\}) = P(\{X \neq Y\}) = 0.$$

Now, if $X, Y$ were jointly continuous with joint density $f_{X,Y}$, then

$$\begin{aligned} &P(\{(X,Y) \notin \Delta\}) = P(\{X < Y\}) + P(\{X > Y\}) \\ &= \int_{-\infty}^{+\infty} \int_{-\infty}^{y} f_{X,Y}(x,y)\,\mathrm{d}x\,\mathrm{d}y + \int_{-\infty}^{+\infty} \int_{y}^{+\infty} f_{X,Y}(x,y)\,\mathrm{d}x\,\mathrm{d}y \\ &= \int_{-\infty}^{+\infty} \int_{-\infty}^{+\infty} f_{X,Y}(x,y)\,\mathrm{d}x\,\mathrm{d}y = P_{X,Y}(\mathbb{R}^2) = 1, \end{aligned}$$

which is a contradiction.

**8.67** Using the substitution $u = x + y$ and $v = x - y$, we find that

$$\begin{aligned} P_{U,V}((a,b] \times (c,d]) &= P(\{X+Y \in (a,b], X-Y \in (c,d]\}) \\ &= \iint_{\substack{x+y \in (a,b] \\ x-y \in (c,d]}} f_{X,Y}(x,y)\,\mathrm{d}x\,\mathrm{d}y \\ &= \int_c^d \int_a^b \tfrac{1}{2} f_{X,Y}\left(\tfrac{u+v}{2}, \tfrac{u-v}{2}\right) \mathrm{d}u\,\mathrm{d}v \end{aligned}$$

for any $a \leqslant b$ and $c \leqslant d$. It follows that $U$ and $V$ are jointly continuous with joint density $f_{U,V}(u,v) = \frac{1}{2} f_{X,Y}\left(\frac{u+v}{2}, \frac{u-v}{2}\right)$.

**8.68** The "if" part is obvious because every interval is a Borel set.

To prove the "only if" part, consider the family $\mathcal{A}$ consisting of all Borel sets $A \subset \mathbb{R}$ such that the events $\{X \in A\}$ and $\{Y \leqslant b\}$ are independent for every $b \in \mathbb{R}$, and the family $\mathcal{B}$ consisting of all Borel sets $B \subset \mathbb{R}$ such that the events $\{X \in A\}$ and $\{Y \in B\}$ are independent for every $A \in \mathcal{A}$. We claim that both $\mathcal{A}$ and $\mathcal{B}$ in fact contain *all* Borel sets. If so, then it follows immediately that $\{X \in A\}$ and $\{Y \in B\}$ are independent for any Borel sets $A, B \subset \mathbb{R}$.

Let us verify the claim to complete the proof. It is not hard to see that both $\mathcal{A}$ and $\mathcal{B}$ are $\sigma$-fields (refer to Chapter 5). Since $X$ and $Y$ are independent, $\mathcal{A}$ contains all intervals of the form $(-\infty, x)$. But such intervals generate the $\sigma$-field of Borel sets, so $\mathcal{A}$ must indeed contain all Borel sets. This, in turn, implies that $\mathcal{B}$ contains all intervals of the form $(-\infty, x)$, and so it must also contain all Borel sets. This proves the claim.

**8.69** For any Borel sets $A, B$ and any Borel functions $f, g$, the inverse images $f^{-1}(A)$ and $g^{-1}(B)$ are also Borel sets, so the events $\{f(X) \in A\} = \{X \in f^{-1}(A)\}$ and $\{g(Y) \in B\} = \{Y \in g^{-1}(B)\}$ are independent by Problem 8.68. This means that $f(X)$ and $g(Y)$ are independent random variables.

**8.70** Observe that for every $a, b \in \mathbb{R}$,

$$P(\{X \leqslant a\})P(\{Y \leqslant b\}) = \int_{-\infty}^{a} f_X(x)\,\mathrm{d}x \int_{-\infty}^{b} f_Y(y)\,\mathrm{d}y$$
$$= \int_{-\infty}^{b}\int_{-\infty}^{a} f_X(x) f_Y(y)\,\mathrm{d}x\,\mathrm{d}y,$$
$$P(\{X \leqslant a\} \cap \{Y \leqslant b\}) = \int_{-\infty}^{b}\int_{-\infty}^{a} f_{X,Y}(x,y)\,\mathrm{d}x\,\mathrm{d}y.$$

It follows that $P(\{X \leqslant a\} \cap \{Y \leqslant b\}) = P(\{X \leqslant a\})P(\{Y \leqslant b\})$ if and only if $f_{X,Y}(x,y) = f_X(x) f_Y(y)$, which is what we want to prove.

**8.71** a) The value of $c$ can be found from

$$1 = P_{X,Y}(\mathbb{R}^2) = \int_{-\infty}^{+\infty}\int_{-\infty}^{+\infty} f_{X,Y}(x,y)\,\mathrm{d}x\,\mathrm{d}y$$
$$= \int_{-\infty}^{0}\int_{-\infty}^{0} c e^{x+y}\,\mathrm{d}x\,\mathrm{d}y = c\left(\int_{-\infty}^{0} e^x\,\mathrm{d}x\right)^2 = c.$$

b) The probability that $X < Y$ is

$$P(\{X < Y\}) = P_{X,Y}(\{(x,y) \in \mathbb{R}^2 : x < y\})$$
$$= \int_{-\infty}^{+\infty}\int_{-\infty}^{y} f_{X,Y}(x,y)\,\mathrm{d}x\,\mathrm{d}y = \int_{-\infty}^{0}\int_{-\infty}^{y} e^{x+y}\,\mathrm{d}x\,\mathrm{d}y$$
$$= \int_{-\infty}^{0} e^y \int_{-\infty}^{y} e^x\,\mathrm{d}x\,\mathrm{d}y = \int_{-\infty}^{0} e^{2y}\,\mathrm{d}y = \frac{1}{2}.$$

c) By (8.3) and (8.4),

$$f_X(x) = \int_{-\infty}^{+\infty} f_{X,Y}(x,y)\,\mathrm{d}y = \int_{-\infty}^{0} e^{x+y}\,\mathrm{d}y = e^x \quad \text{if } x \leqslant 0,$$
$$f_Y(y) = \int_{-\infty}^{+\infty} f_{X,Y}(x,y)\,\mathrm{d}x = \int_{-\infty}^{0} e^{x+y}\,\mathrm{d}x = e^y \quad \text{if } y \leqslant 0.$$

Otherwise, $f_X(x) = f_Y(y) = 0$.
d) If $x, y \leqslant 0$, then

$$f_{X,Y}(x,y) = e^{x+y} = e^x e^y = f_X(x) f_Y(y).$$

If $x > 0$ or $y > 0$, then

$$f_{X,Y}(x,y) = 0 = f_X(x) f_Y(y).$$

By Problem 8.70, it follows that $X$ and $Y$ are independent.

**8.72** Since $X$ and $Y$ are independent, it follows that for any integer $k$ between 0 and $m+n$,

$$\begin{aligned}
&P(\{X+Y=k\}) \\
&= \sum_{i=0}^{k} P(\{X=i\} \cap \{Y=k-i\}) \\
&= \sum_{i=0}^{k} P(\{X=i\})P(\{Y=k-i\}) \\
&= \sum_{i=0}^{k} \binom{m}{i} p^i (1-p)^{m-i} \binom{n}{k-i} p^{k-i} (1-p)^{n-(k-i)} \\
&= \sum_{i=0}^{k} \binom{m}{i} \binom{n}{k-i} p^k (1-p)^{m+n-k} \\
&= \binom{m+n}{k} p^k (1-p)^{m+n-k}.
\end{aligned}$$

The last equality follows by the Van der Monde formula in Problem 2.7. It means that $X+Y$ has the binomial distribution $B(m+n,p)$.

**8.73** Since $X$ and $Y$ are independent, it follows that for any integer $n = 0,1,2,\ldots,$

$$\begin{aligned}
P(\{X+Y=n\}) &= \sum_{k=0}^{n} P(\{X=k\} \cap \{Y=n-k\}) \\
&= \sum_{k=0}^{n} P(\{X=k\})P(\{Y=n-k\}) \\
&= \sum_{k=0}^{n} e^{-\lambda} \frac{\lambda^k}{k!} e^{-\mu} \frac{\mu^{n-k}}{(n-k)!} \\
&= e^{-(\lambda+\mu)} \sum_{k=0}^{n} \frac{\lambda^k}{k!} \frac{\mu^{n-k}}{(n-k)!}.
\end{aligned}$$

According to the binomial formula,

$$\sum_{k=0}^{n} \frac{\lambda^k}{k!} \frac{\mu^{n-k}}{(n-k)!} = \frac{(\lambda+\mu)^n}{n!}.$$

It follows that

$$P(\{X+Y=n\}) = e^{-(\lambda+\mu)} \frac{(\lambda+\mu)^n}{n!},$$

so $X+Y$ has the Poisson distribution with parameter $\lambda+\mu$.

**8.74** Since $X$ and $Y$ are independent random variables, $f_{X,Y}(x,y) = f_X(x)f_Y(y)$; see Problem 8.70. It follows that

$$\begin{aligned}
F_{X+Y}(z) &= \iint_{x+y\leqslant z} f_{X,Y}(x,y)\,dx\,dy \\
&= \iint_{x+y\leqslant z} f_X(x)f_Y(y)\,dx\,dy \\
&= \int_{-\infty}^{+\infty}\int_{-\infty}^{z-x} f_X(x)f_Y(y)\,dx\,dy
\end{aligned}$$

and

$$f_{X+Y}(z) = \frac{dF_{X+Y}}{dz}(z) = \int_{-\infty}^{+\infty} f_X(x)f_Y(z-x)\,dx.$$

# 9

# Expectation and Variance

## 9.1 Theory and Problems

Suppose that you are going to play a simple game in which a die is rolled and you win the amount shown. How much would you be prepared to pay to enter the game? The majority of people agree that a fair amount to pay should be equal to the amount you expect to win *on average*. The following definition gives a precise meaning of the average or mean value of a random variable with finitely many values.

► **Definition 9.1** A random variable $X$ is called *simple* if there is a finite set $C \subset \mathbb{R}$ such that $P(\{X \in C\}) = 1$. The elements of $C$ are called the values of $X$.

For a simple random variable $X$ with $n$ different values $x_1, x_2, \ldots, x_n$, the *expectation* of $X$ is then defined by

$$E(X) = \sum_{i=1}^{n} x_i P(\{X = x_i\}).$$

The expectation $E(X)$ is also referred to as the *mean* or *average value* of $X$.

It is often convenient to express the expectation in terms of the *mass function*

$$f(x) = P(\{X = x\});$$

namely if $X$ is a simple random variable with $n$ different values $x_1, \ldots, x_n$, then

$$E(X) = \sum_{i=1}^{n} x_i f(x_i).$$

**9.1** Let $X$ be the number shown when a single die is rolled (the amount won in the game above). Find $E(X)$.

**9.2** What is the expectation of a random variable with Bernoulli distribution?

**9.3** A friend of yours forgot the 8-digit password necessary to log into his computer. If he tries all possible passwords completely at random, discarding the unsuccessful ones, what is the expected number of attempts needed to find the correct password?

**9.4** A die is rolled twice. What is the expectation of the sum of outcomes?

**9.5** What is the expectation of a random variable with binomial distribution?

▶ **9.6** *Linearity.* Show that $E(X+Y) = E(X) + E(Y)$ and $E(cX) = cE(X)$ for any simple random variables $X$ and $Y$ and any $c \in \mathbb{R}$.

▶ **9.7** *Positivity.* Show that if $X$ is a simple random variable and $X \geqslant 0$, then $E(X) \geqslant 0$.

**9.8** *Monotonicity.* If $X$ and $Y$ are simple random variables such that $X \leqslant Y$, then $E(X) \leqslant E(Y)$.

**9.9** Show that $|E(X)| \leqslant E(|X|)$ for any simple random variable.

**9.10** Let $X$ be a simple random variable such that $X \geqslant 0$. Show that if $E(X) = 0$, then $P(\{X = 0\}) = 1$.

▶ **9.11** Let $X$ be a simple random variable with $n$ different values $x_1, \ldots, x_n$. Show that for any function $g : \mathbb{R} \to \mathbb{R}$,

$$E(g(X)) = \sum_{i=1}^{n} g(x_i) f(x_i),$$

where $f(x) = P(\{X = x\})$ is the mass function.

▶ **9.12** Let $X$ and $Y$ be simple random variables. Verify the *Schwarz inequality*

$$[E(XY)]^2 \leqslant E(X^2) E(Y^2).$$

**9.13** For simple random variables $X$ and $Y$, prove the *triangle inequality*

$$\sqrt{E((X+Y)^2)} \leqslant \sqrt{E(X^2)} + \sqrt{E(Y^2)}.$$

A function $h : (a,b) \to \mathbb{R}$ is called *convex* if $h(px+qy) \leqslant ph(x) + qh(y)$ for all $x, y \in (a,b)$ and for all $p, q \geqslant 0$ such that $p+q=1$. For example, $e^x$ and $|x|$ are convex functions.

▶ **9.14** Prove the following *Jensen inequality*: If $X$ is a simple random variable with values in an interval $(a,b)$ and $h : (a,b) \to \mathbb{R}$ is a convex function, then

$$h(E(X)) \leqslant E(h(X)).$$

**9.15** Show that $E(\mathbb{I}_A) = P(A)$ for any event $A$.

**9.16** Verify the identity $1 - \mathbb{I}_{A\cup B} = (1-\mathbb{I}_A)(1-\mathbb{I}_B)$ and use it to prove that $P(A \cup B) = P(A) + P(B) - P(A \cap B)$.

**9.17** Verify the identity

$$1 - \mathbb{I}_{A_1 \cup A_2 \cup \cdots \cup A_n} = (1-\mathbb{I}_{A_1})(1-\mathbb{I}_{A_2}) \cdots (1-\mathbb{I}_{A_n})$$

and use it to prove the inclusion-exclusion formula stated in Solution 4.20.

**9.18** For any independent simple random variables $X$ and $Y$, show that

$$E(XY) = E(X)E(Y).$$

▶ **Definition 9.2** Let $X$ be a simple random variable with $n$ different values $x_1, \ldots, x_n$. Then, the *variance* of $X$ is defined by

$$V(X) = \sum_{i=1}^{n} (x_i - \mu)^2 P(\{X = x_i\}),$$

where $\mu = E(X)$ is the expectation of $X$.

The variance can also be expressed in terms of the mass function,

$$V(X) = \sum_{i=1}^{n} (x_i - \mu)^2 f(x_i).$$

▶ **9.19** Show that for any simple random variable $X$,

$$V(X) = E([X - E(X)]^2) = E(X^2) - E(X)^2.$$

**9.20** Compute the variance of a random variable with the Bernoulli distribution.

**9.21** Show that if $V(X) = 0$, then $P(\{X = a\}) = 1$ for some $a \in \mathbb{R}$.

**9.22** Show that $V(X + a) = V(X)$ for any $a \in \mathbb{R}$.

**9.23** Show that $V(bX + a) = b^2 V(X)$ for any $a, b \in \mathbb{R}$.

**9.24** Show that

$$V(X + Y) = V(X) + V(Y)$$

if $X$ and $Y$ are independent simple random variables.

**9.25** Compute the variance of a random variable with the binomial distribution.

The above definitions of expectation and variance can readily be extended to the case of discrete random variables. However, the sum may turn into an infinite series and it becomes important to ensure that it converges independently of the order of terms. This is why the assumption about absolute convergence appears below.

We recall (see Definition 8.4) that a discrete random variable $X$ has a countable set of values $C \subset \mathbb{R}$ such that $P(\{X \in C\}) = 1$. The elements of $C$ are called the values of $X$.

▶ **Definition 9.3** Let $X$ be a discrete random variable with different values $x_1, x_2, \ldots$. Then, the *expectation* of $X$ is defined by

$$E(X) = \sum_{i=1}^{\infty} x_i P(\{X = x_i\}),$$

provided that the series is absolutely convergent,

$$\sum_{i=1}^{\infty} |x_i| \, P(\{X = x_i\}) < \infty.$$

If the series in Definition 9.3 fails to be absolutely convergent, then the expectation is undefined.

The expectation of a discrete random variable can be conveniently expressed in terms of the mass function $f(x) = P(\{X = x\})$; namely

$$E(X) = \sum_{i=1}^{\infty} x_i f(x_i),$$

where $x_1, x_2, \ldots$ are the different values of $X$. The last formula is valid whenever the series is absolutely convergent,

$$\sum_{i=1}^{\infty} |x_i| \, f(x_i) < \infty.$$

**9.26** Find an example of a discrete random variable whose expectation is undefined.

**9.27** A biased coin with probability of heads $p$ and tails $1-p$ is tossed repeatedly. Let $X$ be the number of tosses until heads appear for the first time. Compute the expectation of $X$. (The distribution of $X$ is called the *geometric distribution*.)

**9.28** *St. Petersburg problem.* A coin is tossed repeatedly until a tail is obtained and you win $a^k$ dollars, where $k$ is the number of heads before the first tail and $a > 0$. How much would you consider a fair amount to pay to play this game?

**9.29** What is the expectation of a random variable with Poisson distribution?

**9.30** Compute the expectation of a random variable with negative binomial distribution.

**9.31** Let $X$ be a discrete random variable. Show that for any Borel function $g : \mathbb{R} \to \mathbb{R}$, the random variable $g(X)$ is also discrete and, provided that the series below is absolutely convergent,

$$E(g(X)) = \sum_{i=1}^{\infty} g(x_i) f(x_i),$$

where the $x_1, x_2, \ldots$ are the values of $X$ such that $x_i \neq x_j$ if $i \neq j$, and where $f(x) = P(\{X = x\})$ is the mass function of $X$.

**9.32** *Monotone convergence.* Let $X \geqslant 0$ be a discrete random variable and let $0 \leqslant X_1 \leqslant X_2 \leqslant \cdots$ be a non-decreasing sequence of discrete random variables such that $X_n \nearrow X$ as $n \to \infty$. Show that $E(X_n) \nearrow E(X)$ as $n \to \infty$.

▶ **Definition 9.4** Let $X$ be a discrete random variable with distinct values $x_1, x_2, \ldots$ such that the expectation $\mu = E(X)$ is well defined. Then, the *variance* of $X$ is defined by

$$V(X) = \sum_{i=1}^{\infty} (x_i - \mu)^2 P(\{X = x_i\}),$$

provided that the series is convergent.

If the series in Definition 9.4 fails to be convergent or the expectation of $X$ does not exist, then the variance is undefined.

The variance can be expressed in terms of the mass function,

$$V(X) = \sum_{i=1}^{\infty}(x_i - \mu)^2 f(x_i),$$

provided that the series is convergent. The variance of a discrete random variable can also be written as

$$V(X) = E([X - E(X)]^2) = E(X^2) - E(X)^2,$$

whenever the expectations on the right-hand side exist.

**9.33** Find an example of a discrete random variable whose expectation exists, but the variance is undefined.

**9.34** Compute the variance of a random variable with the Poisson distribution.

**9.35** Compute the variance of a random variable with the geometric distribution; see Problem 9.27.

In the next definition, we extend the expectation to the case of an absolutely continuous random variable $X$ with density $f_X$ We recall that in this case,

$$P(\{X \in (a, b]\}) = \int_a^b f_X(x)\, \mathrm{d}x$$

for any $a \leqslant b$; see Definition 8.5.

► **Definition 9.5** The *expectation* of an absolutely continuous random variable $X$ with density $f_X$ is defined by

$$E(X) = \int_{-\infty}^{+\infty} x f_X(x)\, \mathrm{d}x,$$

provided that the integral is absolutely convergent,

$$\int_{-\infty}^{+\infty} |x|\, f_X(x)\, \mathrm{d}x < \infty.$$

If the integral in Definition 9.5 fails to be absolutely convergent, then the expectation is undefined.

**9.36** Find an example of an absolutely continuous random variable $X$ whose expectation is undefined.

Compute $E(X)$ in the following problems:

**9.37** $X$ has the uniform distribution on $[a, b]$ with density $f_X(x) = \frac{1}{b-a}$ if $x \in [a, b]$ and 0 if $x \notin [a, b]$.

**9.38** $X$ has the *exponential distribution* with density

$$f_X(x) = \begin{cases} 0 & \text{for } x < 0 \\ \lambda e^{-\lambda x} & \text{for } x \geqslant 0. \end{cases}$$

**9.39** $X$ has the *Cauchy distribution* with density

$$f_X(x) = \frac{1}{\pi(1 + x^2)}.$$

**9.40** $X$ has the *normal distribution* with density

$$f_X(x) = \frac{1}{\sqrt{2\pi b^2}} \exp\left(-\frac{(x-a)^2}{2b^2}\right),$$

where $a \in \mathbb{R}$ and $b > 0$ are given numbers.

The following theorem is an analog of the properties in Problems 9.11 and 9.31. It is often referred to as the *law of the unconscious statistician.* (The reason for this curious name is that users of the formula below sometimes treat it as if it were the definition of expectation, being unconscious of the actual Definition 9.5.)

▶ **Theorem 9.1** *Let $X$ be an absolutely continuous random variable with density $f_X$ and let $Y = g(X)$. Then,*

$$E(Y) = \int_{-\infty}^{+\infty} g(x) f_X(x)\, dx,$$

*provided that the integral is absolutely convergent, i.e.,*

$$\int_{-\infty}^{+\infty} |g(x)|\, f_X(x)\, dx < \infty.$$

**9.41** Use Theorem 9.1 to compute the expectation of a random variable with log normal distribution (see Problem 8.55 for the definition of log normal distribution).

▶ **Definition 9.6** Let $X$ be an absolutely continuous random variable with density $f_X$ such that the expectation $\mu = E(X)$ is well defined. Then, the *variance* of $X$ is defined by

$$V(X) = \int_{-\infty}^{+\infty} (x - \mu)^2 f_X(x)\, dx,$$

provided that the integral is convergent.

If the integral fails to be convergent or the expectation $\mu$ does not exist, then the variance is undefined.

**9.42** Use Theorem 9.1 to verify that

$$V(X) = E(X^2) - E(X)^2$$

for any absolutely continuous random variable $X$, provided that both expectations on the right-hand side exist.

**9.43** Compute the variance of a random variable with exponential distribution.

**9.44** Compute the variance of a random variable with normal distribution.

It is sometimes necessary to use the expectation of a random variable $X$ which is a mixture of discrete and absolutely continuous distributions. By this, we mean that $X$ can be written as a linear combination $X = aY + bZ$, where $a, b \in \mathbb{R}$, $Y$ is a discrete and $Z$ an absolutely continuous random variable. Then, the expectation of $X$ can be defined by

$$E(X) = aE(Y) + bE(Z),$$

provided that both $E(Y)$ and $E(Z)$ exist.

Find the expectation $E(X)$ of a random variable whose distribution $P_X$ is defined as follows for any Borel subset $A$ of $\mathbb{R}$:

**9.45** $P_X(A) = \frac{1}{3}\delta_0(A) + \frac{2}{3}P_2(A)$, where $P_2$ is an absolutely continuous probability measure with density $f(x) = \frac{1}{2}\mathbb{I}_{[1,3]}$ (see Problem 6.39).

∘ **9.46** $P_X(A) = \frac{1}{8}\delta_1(A) + \frac{3}{8}\delta_2(A) + \frac{1}{2}P_3(A)$, where $P_3$ is an absolutely continuous probability measure with density $f(x) = \frac{1}{2}\mathbb{I}_{[1,3]}$ (see Problem 6.40).

∘ **9.47** $P_X(A) = \frac{1}{4}\delta_0(A) + \frac{1}{2}\delta_2(A) + \frac{1}{4}P_3(A)$, where $P_3$ is an absolutely continuous probability measure with density $f(x) = e^{-x}\mathbb{I}_{[0,+\infty)}$ (see Problem 6.41).

**9.48** On your way to work you have to drive through a busy junction, where you may be stopped at traffic lights. The cycle of the traffic lights is 2 minutes of green followed by 3 minutes of red. What is the expected delay in the journey if you arrive at the junction at a random time uniformly distributed over the whole 5-minute cycle?

So far, we have considered the expectation of random variables with discrete or absolutely continuous distribution and their mixtures. These are the most common random variables in practice. However, by no means do they exhaust all possibilities. There is another sort of random variable called *singular*, which is largely beyond the scope of this course (but see Problems 9.49 and 9.50 for an example).

$*$ **9.49** A fair coin is tossed repeatedly. For each occurrence of heads, you win $\frac{2}{3^k}$, where $k$ is the number of the toss. For tails, you win nothing. Show that the gain $X$ after infinitely many tosses is a singular random variable, i.e., neither discrete nor absolutely continuous nor a mixture thereof.

For an arbitrary non-negative random variable $X$, the expectation can be defined in a similar manner as the Lebesgue integral for non-negative functions, cf. Definition 6.3. The assumption that $X$ is non-negative can then be relaxed. The summability condition below plays the same role in preventing infinities as absolute convergence does in the case of discrete and absolutely continuous random variables (for which summability and absolute convergence are equivalent).

▶ **Definition 9.7** The *expectation* is a mapping

$$X \mapsto E(X) \in [0, \infty]$$

defined for all non-negative random variables $X$ that satisfies the following conditions:

(a) $E(\mathbb{I}_A) = P(A)$,

(b) (*linearity*) $E(X+Y) = E(X) + E(Y)$ and $E(cX) = cE(X)$,

(c) (*monotone convergence*) if $X_n \nearrow X$, then $E(X_n) \nearrow E(X)$

for any event $A$, any $c \geqslant 0$, and any non-negative random variables $X$, $Y$, and $X_1, X_2, \ldots$ . In (c), $X_n \nearrow X$ means that $X_n$ is a non-decreasing sequence convergent to $X$ with probability 1.

These conditions generalize the properties of discrete and absolutely continuous random variables discussed earlier in this chapter. In particular, condition (c) is an extension of monotone convergence for discrete random variables established in Problem 9.32. Condition (c) is also an extension of an analogous property of continuous random variables. In the latter case, the expectation can be expressed as the Lebesgue integral of a density function, satisfying condition (c) of Definition 6.3.

▶ **Definition 9.8** We say that a non-negative random variable $X \geqslant 0$ is *summable* if the expectation of $X$ is finite, $E(X) < \infty$. For an arbitrary random variable $X$, we put

$$X^+ = \begin{cases} X & \text{if } X \geqslant 0 \\ 0 & \text{if } X < 0, \end{cases} \qquad X^- = \begin{cases} 0 & \text{if } X \geqslant 0 \\ -X & \text{if } X < 0, \end{cases}$$

which are known as the *positive* and *negative parts* of $X$. Clearly, $X = X^+ - X^-$. We say that $X$ itself is *summable* whenever the two non-negative random variables $X^+$ and $X^-$ are summable.

For a summable random variable $X$ with positive and negative parts $X^+$ and $X^-$, the *expectation* is defined by

$$E(X) = E(X^+) - E(X^-).$$

The reader familiar with integration theory will readily recognize that this is the definition of the *integral of $X$ with respect to the probability measure $P$*, so we can write

$$E(X) = \int_\Omega X(\omega)\, P(\mathrm{d}\omega).$$

**$*$ 9.50** What is the expected gain $E(X)$ after infinitely many tosses of the coin in Problem 9.49?

## 9.2 Hints

**9.1** List all the values of $X$ and their probabilities. You can use the mass function to express the expectation.

**9.2** What are the values of a random variable with Bernoulli distribution? What are the corresponding probabilities?

**9.3** What is the probability that the password is found at the first attempt? What is the probability that it is not found at the first attempt? What is the probability that it is found at the second attempt? What is the probability that it is not found at the second attempt? ... What is the probability that the password is found at the $k$th attempt?

Now you can use the mass function to compute the expectation.

**9.4** List all possible values of the sum of outcomes. What are the corresponding probabilities? Now, multiply the values by the probabilities and add up to compute the expectation.

Another way is to use Problem 9.6 and the expectation of the outcome of a single roll of a die found in Problem 9.1.

**9.5** It is not hard to compute the expectation directly from Definition 9.1 or using the mass function.

Alternatively, an elegant argument can be obtained by expressing a binomial random variable as a sum of Bernoulli random variables and using Problem 9.6.

**9.6** Each value of $X + Y$ is the sum of a value of $X$ and a value of $Y$. Each value of $cX$ is equal to $c$ times a value of $X$.

**9.7** The values $x_i$ of $X$ must be non-negative.

**9.8** Consider $Y - X$ and use linearity and positivity.

**9.9** Apply Problem 9.8 to the inequalities $X \leqslant |X|$ and $-X \leqslant |X|$.

**9.10** Do all the values $x_i$ of $X$ have to be zero if $X \geqslant 0$ and $E(X) = 0$? If not, what can you say about the set $\{X = x_i\}$?

**9.11** The random variable $g(X)$ is constant and equal to $g(x_i)$ on the set $\{X = x_i\}$. However, it may happen that $g(x_i) = g(x_j)$, even though $x_i \neq x_j$.

**9.12** Consider the quadratic function $f(t) = E\left[(tX - Y)^2\right]$. How many real roots can it have?

**9.13** Taking the square of either side of the triangle inequality, show that it reduces to the Schwarz inequality.

**9.14** If $h$ is a convex function on $(a, b)$ and $x_1, \ldots, x_n \in (a, b)$, then $h(p_1x_1 + \cdots + p_nx_n) \leqslant p_1h(x_1) + \cdots + p_nh(x_n)$ for any $p_1, \ldots, p_n \geqslant 0$ such that $p_1 + \cdots + p_n = 1$.

**9.15** This is immediate from Definition 9.1.

**9.16** The identities $\mathbb{I}_{\Omega \setminus A} = 1 - \mathbb{I}_A$ and $\mathbb{I}_{A \cap B} = \mathbb{I}_A \mathbb{I}_B$ may be useful.

**9.17** Extend Solution 9.16 to $n$ sets.

**9.18** The values of $XY$ are of the form $xy$, where $x$ is a value of $X$ and $y$ is a value of $Y$.

**9.19** Use Problem 9.11 with $g(x) = (x - \mu)^2$.

**9.20** What is the mass function of a random variable with the Bernoulli distribution?

**9.21** Use Problem 9.10.

**9.22** Observe that $E(X+a) = E(X)+a$ and use the definition of variance.

**9.23** You already know that $E(bX) = bE(X)$ and $V(X + a) = V(X)$.

**9.24** Use the fact that $E(XY) = E(X)E(Y)$ if $X$ and $Y$ are independent.

**9.25** Consider the sum of $n$ independent random variables, each having the Bernoulli distribution.

**9.26** Take any sequence of positive numbers $p_1, p_2, \ldots$ such that $\sum_{i=1}^{\infty} p_i = 1$ and find a sequence of positive numbers $x_1, x_2, \ldots$ such that the series $\sum_{i=1}^{\infty} x_i p_i$ is divergent. Use the $x_i$ as the values of the random variable with the corresponding probabilities $p_i$.

**9.27** The event $X = k$ means that the first $k-1$ tosses produce heads, followed by tails at toss $k$. What is the probability of this event? To compute the expectation, you will need to find the sum of a certain series. The Taylor expansion of $\frac{1}{(1-x)^2}$ around 0 may help.

**9.28** What is the probability that the number of heads before the first tail is $k$? What is the expectation of the amount won in this game? Have you checked that the series defining the expectation is absolutely convergent?

**9.29** A Poisson random variable is discrete with values in $\{0, 1, 2 \ldots\}$. What are the probabilities $P(\{X = k\})$? Now use Definition 9.3 to compute the expectation. You will need the Taylor expansion of $e^x$ about $x = 0$.

**9.30** You may need the Taylor expansion of $\frac{1}{(1-x)^{k+1}}$ about $x = 0$.

**9.31** If $x_1, x_2, \ldots$ are the values of $X$, then $g(x_1), g(x_2), \ldots$ are those of $g(X)$. However, the $g(x_i)$ do not have to be different from one another even if the $x_i$ are.

**9.32** It is not hard to show that $\lim_{n\to\infty} E(X_n) \leqslant E(X)$. To obtain the reverse inequality, try to prove that $\lim_{n\to\infty} E(X_n) \geqslant E(X) - \varepsilon$ for any $\varepsilon > 0$ by showing that for $n$ large enough, $X_n \geqslant X - \varepsilon$ on a set of probability arbitrarily close to 1. This is easier if $X$ is a bounded random variable and you may want to begin with this case.

**9.33** Find two numbers $a, b > 0$ such that the geometric series $\sum_{k=0}^{\infty} b^k$ and $\sum_{k=0}^{\infty} a^k b^k$ are convergent, but $\sum_{k=0}^{\infty} a^{2k} b^k$ is divergent.

**9.34** First compute $E(X^2)$.

**9.35** First compute $E(X^2)$. The Taylor expansions of $\frac{1}{(1-x)^2}$ and $\frac{1}{(1-x)^3}$ around 0 may help.

**9.36** The problem amounts to specifying a density $f_X$ such that the integral $\int_{-\infty}^{\infty} |x| \, f_X(x) \, dx$ is divergent.

**9.37** What is the density of a random variable with uniform distribution? Use Definition 9.5 to find the expectation. Is the integral absolutely convergent?

**9.38** Use Definition 9.5. Is the integral absolutely convergent?

**9.39** Use Definition 9.5. Is the integral absolutely convergent?

**9.40** Use Definition 9.5. Is the integral absolutely convergent?

**9.41** Use $g(x) = e^x$ in Theorem 9.1.

**9.42** Use Theorem 9.1 with $g(x) = x^2$ to express $E(X^2)$.

**9.43** The density is given in Problem 9.38.

**9.44** $\int_{-\infty}^{+\infty} y^2 \exp(-y^2/2)\, dy = \sqrt{2\pi}$.

**9.45** Write $X$ as a linear combination of a discrete and an absolutely continuous random variable.

**9.46** See Hint 9.45.

**9.47** See Hint 9.45.

**9.48** The delay is a random variable. Find its distribution. Is it a mixture of discrete and absolutely continuous distributions?

**9.49** The values of $X$ are in the Cantor set of Problem 5.23.

**9.50** Even though $X$ is a singular random variable (see Problem 9.49), finding $E(X)$ is not hard. Let $X_n$ be the total gain after $n$ tosses of the coin. Observe that $X_n \nearrow X$, so $E(X_n) \nearrow E(X)$. Find the expectation of the $X_n$, which are discrete random variables. Then, take the limit.

## 9.3 Solutions

**9.1** The values of $X$ are the outcomes 1, 2, 3, 4, 5 or 6 of rolling a die, each having probability $\frac{1}{6}$. It follows that

$$E(X) = \sum_{i=1}^{6} i f(i) = \sum_{i=1}^{6} \frac{i}{6} = \frac{1+2+3+4+5+6}{6} = \frac{7}{2}.$$

**9.2** If $P(\{X = 1\}) = p$ and $P(\{X = 0\}) = 1 - p$, then

$$E(X) = 1 \times p + 0 \times (1 - p) = p.$$

**9.3** Let $n = 10^8$ be the number of all possible passwords and let $X$ be the number of attempts before the correct password is found (including the last successful attempt). The probability that the correct password is found at the $k$th attempt is

$$P(\{X = k\}) = \frac{n-1}{n}\frac{n-2}{n-1}\cdots\frac{n-k+1}{n-k+2}\frac{1}{n-k+1} = \frac{1}{n}.$$

Here, $\frac{n-1}{n}$ is the probability that the password is not found at the first attempt ($n-1$ unsuccessful passwords out of $n$). Given that the password has not been found at the first attempt, $\frac{n-2}{n-1}$ is the probability that it has not been found at the second attempt (there are $n-1$ passwords left, including $n-2$ unsuccessful ones). ... Given that the password has not been found so far, $\frac{n-k+1}{n-k+2}$ is the probability that is is not found at the $(k-1)$st attempt. Finally, $\frac{1}{n-k+1}$ is the probability that the password is found at the $k$th attempt.

It follows that

$$E(X) = \sum_{k=0}^{n} kP(\{X = k\}) = \sum_{k=0}^{n} \frac{k}{n}$$
$$= \frac{n+1}{2} = \frac{100\,000\,001}{2} = 50\,000\,000.5\,.$$

**9.4** Let $X$ and $Y$ be the results of the first and, respectively, the second roll of the die. Both have six values $1, \ldots, 6$. Their sum $Z = X + Y$ has 11 different values, $2, 3, \ldots, 12$. The expectation of $Z$ is

$$E(Z) = \sum_{k=2}^{12} kP(\{Z = k\}) = \sum_{k=2}^{12} k \sum_{i+j=k} P(\{X = i, Y = j\})$$
$$= \sum_{k=2}^{12} \sum_{i+j=k} (i+j)P(\{X = i, Y = j\}) = \sum_{i,j=1}^{6} \frac{i+j}{36} = 7.$$

**9.5** For a random variable $X$ with binomial distribution,

$$P(\{X = k\}) = \frac{n!}{k!\,(n-k)!}p^k\,(1-p)^{n-k}, \quad k = 0, 1, 2, \ldots, n,$$

so

$$E(X) = \sum_{k=0}^{n} kP(\{X = k\}) = \sum_{k=1}^{n} \frac{n!}{(k-1)!\,(n-k)!}p^k\,(1-p)^{n-k}$$
$$= np \sum_{k=1}^{n} \frac{(n-1)!}{(k-1)!\,(n-k)!}p^{k-1}\,(1-p)^{n-k}$$
$$= np \sum_{l=0}^{n-1} \frac{(n-1)!}{l!\,(n-l)!}p^l\,(1-p)^{n-l} = np.$$

**9.6** Suppose that $X$ has $m$ different values $x_1, \ldots, x_m$ and $Y$ has $n$ different values $y_1, \ldots, y_n$. Then, $Z + Y$ has no more than $m + n$ different values $z_1, \ldots, z_k$, where $k \leqslant m + n$, so $X + Y$ is a simple random variable.

We put $A_l = \{(i,j) : x_i + y_j = z_j\}$, $B_i = \{X = x_i\}$, and $C_j = \{Y = y_j\}$. Then,

$$P(\{X + Y = z_l\}) = \sum_{(i,j)\in A_l} P(B_i \cap C_j)$$

and

$$\begin{aligned}
E(X+Y) &= \sum_{l=1}^{k} z_l P(\{X + Y = z_l\}) \\
&= \sum_{l=1}^{k} \sum_{(i,j)\in A_l} z_l P(B_i \cap C_j) \\
&= \sum_{l=1}^{k} \sum_{(i,j)\in A_l} (x_i + y_j) P(B_i \cap C_j) \\
&= \sum_{i=1}^{m} \sum_{j=1}^{n} (x_i + y_j) P(B_i \cap C_j) \\
&= \sum_{i=1}^{m} x_i \sum_{j=1}^{n} P(B_i \cap C_j) + \sum_{j=1}^{n} y_j \sum_{i=1}^{m} P(B_i \cap C_j) \\
&= \sum_{i=1}^{m} x_i P(B_i) + \sum_{j=1}^{n} y_j P(C_j) = E(X) + E(Y).
\end{aligned}$$

If $X$ has $m$ different values $x_1, \ldots, x_m$, then $cX$ also has $m$ different values $cx_1, \ldots, cx_m$ (unless $c = 0$, in which case the equality $E(cX) = cE(X)$ is trivial). Then,

$$E(cX) = \sum_{i=1}^{m} cx_i P(\{cX = cx_i\}) = c \sum_{i=1}^{m} x_i P(\{X = x_i\}) = cE(X).$$

**9.7** Suppose that $X$ has $m$ different values $x_1, \ldots, x_m$. Since $X \geqslant 0$, it follows that $x_i \geqslant 0$ for each $i$ and

$$E(X) = \sum_{i=1}^{m} x_i P(\{X = x_i\}) \geqslant 0.$$

**9.8** If $X$ and $Y$ are simple random variables, then so is $Y - X$. Since $X \leqslant Y$, we have $Y - X \geqslant 0$ and

$$E(Y) - E(X) = E(Y - X) \geqslant 0$$

by Problems 9.6 and 9.7, which proves the required inequality.

**9.9** If $X$ is a simple random variable, then so is $|X|$. Since $X \leqslant |X|$ and $-X \leqslant |X|$,

$$E(X) \leqslant E(|X|) \quad \text{and} \quad -E(X) \leqslant E(|X|)$$

by Problems 9.6 and 9.8. It follows that $|E(X)| \leqslant E(|X|)$.

**9.10** Suppose that $X$ has $m$ different values $x_1, \dots, x_m$. Since $X \geqslant 0$, all the $x_i$ are non-negative. If so, then

$$E(X) = \sum_{i=1}^{n} x_i P(\{X = x_i\}) = 0$$

only if all the terms $x_i P(\{X = x_i\})$ in the sum are zero. This means that either $x_i = 0$ or $P(\{X = x_i\}) = 0$ for each $i = 1, \dots, n$. Taking $A$ to be the union of all sets $\{X = x_i\}$ such that $x_i \neq 0$, we find that $X = 0$ in $\Omega \setminus A$ and $P(A) = 0$, proving the assertion.

**9.11** If $X$ has $n$ different values $x_1, \dots, x_n$, then $g(X)$ has at most $n$ different values $z_1, \dots, z_k$, where $k \leqslant n$. Put

$$A_l = \{i : g(x_i) = z_l\}.$$

Then,

$$P(\{g(X) = z_l\}) = \sum_{i \in A_l} P(\{X = x_i\}) = \sum_{i \in A_l} f(x_i)$$

and

$$\begin{aligned} E(g(X)) &= \sum_{l=1}^{k} z_l P(\{g(X) = z_l\}) = \sum_{l=1}^{k} z_l \sum_{i \in A_l} f(x_i) \\ &= \sum_{l=1}^{k} \sum_{i \in A_l} z_l f(x_i) = \sum_{l=1}^{k} \sum_{i \in A_l} g(x_i) f(x_i) = \sum_{i=1}^{n} g(x_i) f(x_i). \end{aligned}$$

**9.12** The quadratic function

$$f(t) = E\left[(tX - Y)^2\right] = t^2 E(X^2) - 2t E(XY) + E(Y^2)$$

is non-negative, so it has at most one real root and

$$[2E(XY)]^2 - 4E(X^2)E(Y^2) \leqslant 0,$$

which implies the Schwarz inequality.

**9.13** From the Schwarz inequality, we have

$$\begin{aligned} E((X+Y)^2) &= E(X^2) + 2E(XY) + E(Y^2) \\ &\leqslant E(X^2) + 2\sqrt{E(X^2)E(Y^2)} + E(Y^2) \\ &= \left(\sqrt{E(X^2)} + \sqrt{E(Y^2)}\right)^2. \end{aligned}$$

Taking the square root of both sides, we obtain the triangle inequality.

**9.14** If $h : (a,b) \to \mathbb{R}$ is a convex function, then

$$h\left(\sum_{i=1}^{n} p_i x_i\right) \leqslant \sum_{i=1}^{n} p_i h(x_i)$$

for any $x_1, \ldots, x_n \in (a,b)$ and for any $p_1, \ldots, p_n \geqslant 0$ such that $\sum_{i=1}^{n} p_i = 1$. If $X$ has $n$ different vales $x_1, \ldots, x_n$, then

$$E(X) = \sum_{i=1}^{n} x_i f(x_i),$$

where $f(x) = P(\{X = x\})$ is the mass function. Since $\sum_{i=1}^{n} f(x_i) = 1$, we have

$$h(E(X)) = h\left(\sum_{i=1}^{n} x_i f(x_i)\right) \leqslant \sum_{i=1}^{n} f(x_i) h(x_i) = E(h(X))$$

by Problem 9.11.

**9.15** Just substitute $X = \mathbb{I}_A$ in Definition 9.1 to get $E(\mathbb{I}_A) = P(A)$.

**9.16** First, observe that $\mathbb{I}_{\Omega \setminus A} = 1 - \mathbb{I}_A$ and $\mathbb{I}_{A \cap B} = \mathbb{I}_A \mathbb{I}_B$ for any events $A$ and $B$. It follows that

$$\begin{aligned} 1 - \mathbb{I}_{A \cup B} &= \mathbb{I}_{\Omega \setminus (A \cup B)} \\ &= \mathbb{I}_{(\Omega \setminus A) \cap (\Omega \setminus B)} = \mathbb{I}_{\Omega \setminus A} \mathbb{I}_{\Omega \setminus B} = (1 - \mathbb{I}_A)(1 - \mathbb{I}_B). \end{aligned}$$

Here, we have used de Morgan's laws. Expanding the right-hand side, we obtain

$$\mathbb{I}_{A \cup B} = \mathbb{I}_A + \mathbb{I}_B - \mathbb{I}_A \mathbb{I}_B = \mathbb{I}_A + \mathbb{I}_B - \mathbb{I}_{A \cap B}.$$

It remains to take the expectation of both sides to get $P(A \cup B) = P(A) + P(B) - P(A \cap B)$.

**9.17** We can use the identities $\mathbb{I}_{\Omega\setminus A} = 1 - \mathbb{I}_A$ and $\mathbb{I}_{A\cap B} = \mathbb{I}_A\mathbb{I}_B$ along with de Morgan's laws again to get

$$\begin{aligned}1 - \mathbb{I}_{A_1\cup\cdots\cup A_n} &= \mathbb{I}_{\Omega\setminus(A_1\cup\cdots\cup A_n)} = \mathbb{I}_{(\Omega\setminus A_1)\cap\cdots\cap(\Omega\setminus A_n)}\\ &= \mathbb{I}_{\Omega\setminus A_1}\cdots\mathbb{I}_{\Omega\setminus A_n} = (1-\mathbb{I}_{A_1})(1-\mathbb{I}_{A_2})\cdots(1-\mathbb{I}_{A_n}).\end{aligned}$$

Expanding the right hand-side, we obtain

$$\begin{aligned}\mathbb{I}_{A_1\cup\cdots\cup A_n} &= \sum_i \mathbb{I}_{A_i} - \sum_{i,j}\mathbb{I}_{A_i\cap A_j} + \sum_{i,j,k}\mathbb{I}_{A_i\cap A_j\cap A_k} - \cdots\\ &\quad +(-1)^{n+1}\mathbb{I}_{A_1\cap\cdots\cap A_n}.\end{aligned}$$

Finally, we take the expectation of both sides to get the inclusion-exclusion formula stated in Solution 4.20.

**9.18** Suppose that $X$ has $m$ different values $x_1, \dots, x_m$ and $Y$ has $n$ different values $y_1, \dots, y_n$. Then, $XY$ has at most $mn$ different values $z_1, \dots, z_k$, where $k \leqslant mn$. We put $A_l = \{(i,j) : x_i y_j = z_l\}$, $B_i = \{X = x_i\}$, and $C_j = \{Y = y_j\}$. Then, by independence

$$P(\{XY = z_l\}) = \sum_{(i,j)\in A_l} P(\{B_i \cap C_j\}) = \sum_{(i,j)\in A_l} P(B_i)P(C_j).$$

It follows that

$$\begin{aligned}E(XY) &= \sum_{l=1}^{k} z_l P(\{XY = z_l\}) = \sum_{l=1}^{k}\sum_{(i,j)\in A_l} z_l P(B_i)P(C_j)\\ &= \sum_{l=1}^{k}\sum_{(i,j)\in A_l} x_i y_j P(B_i)P(C_j) = \sum_{i=1}^{m}\sum_{j=1}^{n} x_i y_j P(B_i)P(C_j)\\ &= \sum_{i=1}^{m} x_i P(B_i)\sum_{j=1}^{n} y_j P(C_j) = E(X)E(Y).\end{aligned}$$

**9.19** Suppose that $X$ has $n$ different values $x, \dots, x_n$ and mass function $f(x)$. Putting $g(x) = (x-\mu)^2$ with $\mu = E(X)$ in Problem 9.11, we find that

$$E([X - E(X)]^2) = \sum_{i=1}^{n}(x_i - \mu)^2 f(x_i) = V(X).$$

Moreover,

$$\begin{aligned}E([X - E(X)]^2) &= E(X^2 - 2XE(X) + E(X)^2)\\ &= E(X^2) - 2E(X)^2 + E(X)^2 = E(X^2) - E(X)^2.\end{aligned}$$

**9.20** Suppose that $X$ has the Bernoulli distribution $B(1,p)$. Then, the mass function of $X$ is

$$f(0) = 1 - p, \qquad f(1) = p$$

and $\mu = E(X) = p$. It follows that the variance of $X$ is

$$V(X) = (0-p)^2(1-p) + (1-p)^2 p = p(1-p).$$

**9.21** If $V(X) = E([X - E(X)]^2) = 0$, then $P(\{[X - E(X)]^2 = 0\}) = 1$ by Problem 9.10. This, in turn, implies that $P(\{X = E(X)\}) = 1$, completing the proof.

**9.22** Since $E(X+a) = E(X) + a$,

$$V(X+a) = E([X + a - E(X+a)]^2) = E([X - E(X)]^2) = V(x).$$

**9.23** Since $E(bX) = bE(X)$ and $V(bX+a) = V(bX)$ (see above),

$$\begin{aligned} V(bX+a) &= V(bX) = E([bX - E(bX)]^2) \\ &= b^2 E([X - E(X)]^2) = b^2 V(x). \end{aligned}$$

**9.24** If $X$ and $Y$ are independent, then $E(XY) = E(X)E(Y)$ by Problem 9.18. Thus, using Problem 9.19, we obtain

$$\begin{aligned} V(X+Y) &= E([X+Y]^2) - E(X+Y)^2 \\ &= E(X^2) + 2E(XY) + E(Y^2) \\ &\quad -E(X)^2 - 2E(X)E(Y) - E(Y)^2 \\ &= V(X) + V(Y) + 2E(XY) - 2E(X)E(Y) \\ &= V(X) + V(Y). \end{aligned}$$

**9.25** We shall use Problem 9.24. Let $Y_1, Y_2, \ldots$ be a sequence of independent random variables, each having the Bernoulli distribution $B(1,p)$. We know that $V(Y_n) = p(1-p)$ for each $n$; see Solution 9.20. Now, let $X_n = Y_1 + \cdots + Y_n$. By Problem 8.72, $X_n$ has the binomial distribution $B(n,p)$. Now, by Problem 9.24,

$$V(X_n) = V(Y_1) + \cdots + V(Y_n) = np(1-p).$$

Thus, we have computed the variance of any random variable with the binomial distribution $B(n,p)$.

**9.26** Let $X$ be a random variable with values $2, 2^2, 2^3, \ldots$ and corresponding probabilities $P(\{X = 2^i\}) = \frac{1}{2^i}$ for $i = 1, 2, \ldots$ . Because the series

$$\sum_{i=1}^{\infty} 2^i P(\{X = 2^i\}) = \sum_{i=1}^{\infty} 2^i \frac{1}{2^i} = \sum_{i=1}^{\infty} 1$$

is divergent, the expectation of $X$ is undefined.

**9.27** The random variable $X$ takes values $k = 1, 2, \ldots$ with probability

$$P(\{X = k\}) = p(1-p)^{k-1}.$$

(This is called the *geometric distribution*.) The expectation of $X$ is

$$\begin{aligned} E(X) &= \sum_{k=1}^{\infty} kp(1-p)^{k-1} \\ &= p\sum_{k=0}^{\infty}(k+1)(1-p)^k = p\frac{1}{[1-(1-p)]^2} = \frac{1}{p}. \end{aligned}$$

To find the sum of the above series, we can use the Taylor expansion $\frac{1}{(1-x)^2} = \sum_{k=0}^{\infty}(k+1)x^k$.

**9.28** Let $X$ be the number of heads before the first tail. You would expect to win $E(a^X)$ on average, which would be the amount to pay if the game is to be fair. Finding

$$P(\{X = k\}) = \frac{1}{2^{k+1}},$$

we obtain

$$E(a^X) = \sum_{k=0}^{\infty} a^k P(\{X = k\}) = \sum_{k=0}^{\infty} \frac{a^k}{2^{k+1}} = \frac{1}{2-a}$$

if $a < 2$. But if $a \geqslant 2$, then the series is divergent and the expectation is undefined. This is known as the St. Petersburg paradox: no finite amount to pay would make the game fair when $a \geqslant 2$. In this case, since the above series diverges to $\infty$, it is sometimes said that the expected gain would be infinite.

**9.29** By Definition 9.3,

$$\begin{aligned} E(X) &= \sum_{k=0}^{\infty} kP(\{X = k\}) = \sum_{k=0}^{\infty} ke^{-\lambda}\frac{\lambda^k}{k!} \\ &= \lambda e^{-\lambda}\sum_{k=1}^{\infty}\frac{\lambda^{k-1}}{(k-1)!} = \lambda e^{-\lambda}\sum_{k=0}^{\infty}\frac{\lambda^k}{k!} = \lambda e^{-\lambda}e^{\lambda} = \lambda. \end{aligned}$$

**9.30** Let $Y$ be a random variable with negative binomial distribution as in Solution 8.43. Then,

$$\begin{aligned} E(Y) &= \sum_{n=k}^{\infty} nP(\{Y = n\}) = \sum_{n=k}^{\infty} n\binom{n-1}{k-1}p^k(1-p)^{n-k} \\ &= kp^k\sum_{n=k}^{\infty}\binom{n}{k}(1-p)^{n-k}. \end{aligned}$$

To find the sum of this series, we observe that the Taylor expansion of $\frac{1}{(1-x)^{k+1}}$ about 0 is $\sum_{n=k}^{\infty}\binom{n}{k}x^{n-k}$. As a result,

$$E(Y) = \frac{kp^k}{(1-(1-p))^{k+1}} = \frac{kp^k}{p^{k+1}} = \frac{k}{p}.$$

**9.31** Let $X$ be a discrete random variable with distinct values $x_1, x_2, \ldots$ . Let $z_1, z_2, \ldots$ be the different values of $g(X)$. We put $A_l = \{i : g(x_i) = z_l\}$. Then,

$$P(\{g(X) = z_l\}) = \sum_{i \in A_l} z_l f(x_i)$$

and

$$E(g(X)) = \sum_{l=1}^{\infty} z_l P(\{g(X) = z_l\}) = \sum_{l=1}^{\infty} \sum_{i \in A_l} z_l f(x_i)$$
$$= \sum_{l=1}^{\infty} \sum_{i \in A_l} g(x_i) f(x_i) = \sum_{i=1}^{\infty} g(x_i) f(x_i),$$

provided that the series is absolutely convergent.

**9.32** Clearly, $E(X_n)$ is a non-decreasing sequence such that $E(X_n) \leqslant E(X)$, so $\lim_{n\to\infty} E(X_n) \leqslant E(X)$. We need to show that the reverse inequality holds too.

Suppose that $X$ has different values $x_1, x_2, \ldots \geqslant 0$. Let us put $Y_k = X$ and $Y_{k,n} = X_n$ on $\{X = x_1\} \cup \cdots \cup \{X = x_k\}$, and $Y_k = Y_{k,n} = 0$ otherwise. By Definition 9.3,

$$E(X) = \sum_{i=1}^{\infty} x_i P(\{X = x_i\})$$
$$= \lim_{k\to\infty} \sum_{i=1}^{k} x_i P(\{X = x_i\}) = \lim_{k\to\infty} E(Y_k).$$

We claim that for any $\varepsilon > 0$ and any $k = 1, 2, \ldots,$

$$E(X_n) \geqslant E(Y_k) - \varepsilon$$

if $n$ is large enough. If the claim is true, then it follows immediately that $\lim_{n\to\infty} E(X_n) \geqslant E(Y_k)$ for each $k$, which, in turn, implies that $\lim_{n\to\infty} E(X_n) \geqslant E(X)$, as required.

To prove the claim, we take any $\varepsilon > 0$ and put

$$A_n = \{\forall m \geqslant n : X - X_m \leqslant \varepsilon/2\},$$

obtaining an expanding sequence of events $A_1 \subset A_2 \subset \cdots$ such that $P(A_1 \cup A_2 \cup \cdots) = 1$ because $X_n \nearrow X$ as $n \to \infty$. It follows that $P(A_n) \to 1$ as $n \to \infty$. The random variable $Y_k$ has finitely many values, the largest of which we denote by $M$. As a result,

$$\begin{aligned} E(Y_k - X_n) &\leqslant E(Y_k - Y_{k,n}) \\ &= E(\mathbb{I}_{A_n}(Y_k - Y_{k,n})) + E(\mathbb{I}_{\Omega \setminus A_n}(Y_k - Y_{k,n})) \\ &\leqslant E(\mathbb{I}_{A_n}(X - X_n)) + E(\mathbb{I}_{\Omega \setminus A_n} Y_k) \\ &\leqslant P(A_n)\varepsilon/2 + (1 - P(A_n))M \\ &\leqslant \varepsilon/2 + \varepsilon/2 = \varepsilon, \end{aligned}$$

provided that $n$ is large enough, which proves the claim.

**9.33** Let $X$ be a random variable with distribution $P(\{X = \frac{3^k}{2^k}\}) = \frac{1}{2^k}$ for $k = 0, 1, 2, \ldots$ . This discrete distribution is well defined, since

$$\sum_{k=0}^{\infty} \frac{1}{2^k} = 1.$$

The expectation of $X$ is given by the geometric series

$$E(X) = \sum_{k=0}^{\infty} \frac{3^k}{2^k} \frac{1}{2^k} = \sum_{k=0}^{\infty} \left(\frac{3}{4}\right)^k = \frac{1}{1 - \frac{3}{4}} = 4,$$

which converges, since $\frac{3}{4} < 1$. However, trying to compute the expectation of $X^2$, we obtain

$$\sum_{k=0}^{\infty} \left(\frac{3^k}{2^k}\right)^2 \frac{1}{2^k} = \sum_{k=0}^{\infty} \left(\frac{9}{8}\right)^k,$$

which is a divergent geometric series, since $\frac{9}{8} > 1$. It follows that the expectation of $X^2$ is undefined and so is the variance of $X$.

**9.34** Suppose that $X$ has the Poisson distribution with parameter $\lambda$. Then, $E(X) = \lambda$ by Solution 9.29 and

$$\begin{aligned} E(X^2) &= \sum_{k=0}^{\infty} k^2 e^{-\lambda} \frac{\lambda^k}{k!} = e^{-\lambda} \sum_{k=0}^{\infty} [k(k-1) + k] \frac{\lambda^k}{k!} \\ &= \lambda^2 e^{-\lambda} \sum_{k=2}^{\infty} \frac{\lambda^{k-2}}{(k-2)!} + \lambda e^{-\lambda} \sum_{k=1}^{\infty} \frac{\lambda^{k-1}}{(k-1)!} \\ &= \lambda^2 e^{-\lambda} e^{\lambda} + \lambda e^{-\lambda} e^{\lambda} = \lambda^2 + \lambda. \end{aligned}$$

It follows that

$$V(X) = E(X^2) - E(X)^2 = \lambda^2 + \lambda - \lambda^2 = \lambda.$$

**9.35** Let $X$ have the geometric distribution as in Solution 9.27. First, we shall compute the expectation of $X^2$:

$$\begin{aligned} E(X^2) &= \sum_{k=1}^{\infty} k^2 p(1-p)^{k-1} = p\sum_{k=1}^{\infty}[(k+1)k - k](1-p)^{k-1} \\ &= p\sum_{k=0}^{\infty}(k+2)(k+1)(1-p)^k - p\sum_{k=1}^{\infty}(k+1)(1-p)^k \\ &= p\frac{1}{2[1-(1-p)]^3} - p\frac{1}{[1-(1-p)]^2} = \frac{1}{2p^2} - \frac{1}{p}. \end{aligned}$$

To compute the sums of the two series, we use the Taylor expansions $\frac{1}{(1-x)^2} = \sum_{k=0}^{\infty}(k+1)x^k$ and $\frac{1}{(1-x)^3} = \sum_{k=0}^{\infty}\frac{1}{2}(k+2)(k+1)x^k$.
Since, by Solution 9.27, $E(X) = \frac{1}{p}$, it follows that

$$V(X) = E(X^2) - E(X)^2 = \frac{1}{2p^2} - \frac{1}{p} - \frac{1}{p^2} = \frac{1-p}{p^2}.$$

**9.36** Take a random variable $X$ with density

$$f_X(x) = \begin{cases} 0 & \text{for } x < 1 \\ \frac{1}{x^2} & \text{for } x \geqslant 1. \end{cases}$$

This is a well-defined density because $f_X(x) \geqslant 0$ is piecewise continuous and

$$\int_{-\infty}^{\infty} f_X(x)\,\mathrm{d}x = \int_1^{\infty}\frac{1}{x^2}\,\mathrm{d}x = -\frac{1}{x}\Big|_{x=1}^{\infty} = 1.$$

However,

$$\int_{-\infty}^{\infty} |x|\, f_X(x)\,\mathrm{d}x = \int_1^{\infty}\frac{1}{x}\,\mathrm{d}x = \ln x\Big|_{x=1}^{\infty} = \infty,$$

so the expectation of $X$ is undefined.

**9.37** By Definition 9.5,

$$E(X) = \int_{-\infty}^{+\infty} x f_X(x)\,\mathrm{d}x = \int_a^b \frac{x}{b-a}\,\mathrm{d}x = \frac{1}{2}(a+b),$$

the integral being absolutely convergent.

**9.38** Using Definition 9.5 and integrating by parts, we obtain

$$E(X) = \int_{-\infty}^{+\infty} x f_X(x)\,\mathrm{d}x = \int_0^{\infty} x\lambda e^{-\lambda x}\,\mathrm{d}x = \frac{1}{\lambda}.$$

The integral is absolutely convergent.

**9.39** In the case of the Cauchy distribution, the integral in Definition 9.5 fails to be absolutely convergent because

$$\begin{aligned}\int_{-\infty}^{+\infty} |x|\, f_X(x)\, \mathrm{d}x &= \int_{-\infty}^{+\infty} \frac{|x|}{1+x^2}\, \mathrm{d}x \\ &= 2\int_0^{+\infty} \frac{x}{1+x^2}\, \mathrm{d}x = 2\lim_{a\to+\infty}\int_0^a \frac{x}{1+x^2}\, \mathrm{d}x \\ &= \lim_{a\to+\infty} \ln(1+a^2) = \infty,\end{aligned}$$

which means that the expectation is undefined.

**9.40** By Definition 9.5,

$$\begin{aligned}E(X) &= \int_{-\infty}^{+\infty} x\frac{1}{\sqrt{2\pi b^2}}\exp\left(-\frac{(x-a)^2}{2b^2}\right) \mathrm{d}x \\ &= \int_{-\infty}^{+\infty} (by+a)\frac{1}{\sqrt{2\pi}}\exp\left(-\frac{y^2}{2}\right) \mathrm{d}y = a,\end{aligned}$$

since $\int_{-\infty}^{+\infty} \exp(-y^2/2)\, \mathrm{d}y = 1$ and $\int_{-\infty}^{+\infty} y\exp(-y^2/2)\, \mathrm{d}y = 0$, all integrals being absolutely convergent.

**9.41** A random variable $Y$ with log normal distribution can be written as $Y = e^X$, where $X$ has the normal distribution $N(0,1)$. By Theorem 9.1,

$$\begin{aligned}E(Y) &= \int_{-\infty}^{+\infty} e^x f_X(x)\, \mathrm{d}x = \int_{-\infty}^{+\infty} e^x \frac{1}{\sqrt{2\pi}} e^{-\frac{x^2}{2}}\, \mathrm{d}x \\ &= \frac{1}{\sqrt{2\pi e}}\int_{-\infty}^{+\infty} e^{-\frac{(x-1)^2}{2}}\, \mathrm{d}x = \frac{1}{\sqrt{e}}.\end{aligned}$$

**9.42** By Theorem 9.1,

$$E(X^2) = \int_{-\infty}^{+\infty} x^2 f_X(x)\, \mathrm{d}x,$$

so

$$\begin{aligned}V(X) &= \int_{-\infty}^{+\infty} (x-\mu)^2 f_X(x)\, \mathrm{d}x \\ &= \int_{-\infty}^{+\infty} x^2 f_X(x)\, \mathrm{d}x - 2\mu\int_{-\infty}^{+\infty} x f_X(x)\, \mathrm{d}x + \mu^2\int_{-\infty}^{+\infty} f_X(x)\, \mathrm{d}x \\ &= E(X^2) - 2\mu E(X) + \mu^2 = E(X^2) - E(X)^2,\end{aligned}$$

since $\mu = E(X)$.

**9.43** The density of a random variable $X$ with exponential distribution is given in Problem 9.38. Using Theorem 9.1, we compute

$$E(X^2) = \int_{-\infty}^{+\infty} x^2 f_X(x)\,\mathrm{d}x = \int_0^{+\infty} x^2 \lambda e^{-\lambda x}\,\mathrm{d}x = \frac{2}{\lambda^2}.$$

We already know that $E(X) = \frac{1}{\lambda}$; see Solution 9.38. By Problem 9.42, it follows that

$$V(X) = E(X^2) - E(X)^2 = \frac{2}{\lambda^2} - \frac{1}{\lambda^2} = \frac{1}{\lambda^2}.$$

**9.44** Suppose that $X$ has the normal distribution with density

$$f_X(x) = \frac{1}{\sqrt{2\pi b^2}} \exp\left(-\frac{(x-a)^2}{2b^2}\right).$$

By Problem 9.40, $E(X) = a$. Let us compute $V(X) = E\big((X-a)^2\big)$:

$$\begin{aligned} V(X) &= \int_{-\infty}^{+\infty} (x-a)^2 f_X(x)\,\mathrm{d}x \\ &= \frac{1}{\sqrt{2\pi b^2}} \int_{-\infty}^{+\infty} (x-a)^2 \exp\left(-\frac{(x-a)^2}{2b^2}\right)\mathrm{d}x \\ &= \frac{b^2}{\sqrt{2\pi}} \int_{-\infty}^{+\infty} y^2 \exp\left(-\frac{y^2}{2}\right)\mathrm{d}y = b^2, \end{aligned}$$

since $\int_{-\infty}^{+\infty} y^2 \exp(-y^2/2)\,dy = \sqrt{2\pi}$.

**9.45** Let $Y$ be a constant random variable identically equal to 0. It is obviously discrete and $E(Y) = 0$. Let $Z$ be an absolutely continuous random variable with density $f(x) = \frac{1}{2}\mathbb{I}_{[1,3]}$, so $E(Z) = 2$. Then, $X = \frac{1}{3}Y + \frac{2}{3}Z$ and

$$E(X) = \frac{1}{3}E(Y) + \frac{2}{3}E(Z) = \frac{1}{3} \times 0 + \frac{2}{3} \times 2 = \frac{4}{3}.$$

**9.48** Let $X$ be the delay and $T$ the time when you arrive at the junction (counting from the beginning of the current cycle of traffic lights). If $0 \leqslant T < 2$ (green light), then the delay is $X = 0$, and if $2 \leqslant T < 5$ (red light), then $X = 5 - T \in (0, 3]$. Because

$$P(\{0 \leqslant T < 2\}) = \frac{2}{5} \quad \text{and} \quad P(\{2 \leqslant T < 5\}) = \frac{3}{5},$$

the distribution of $X$ can be written as

$$P_X = \frac{2}{5}P_1 + \frac{3}{5}P_2.$$

Here, $P_1 = \delta_0$ and $P_2$ is an absolutely continuous measure with density $f(x) = \frac{1}{3}\mathbb{I}_{(0,3]}$. As in Problem 9.45, we can write $X = \frac{2}{5}Y + \frac{3}{5}Z$, where $Y$ is identically equal to 0 and $Z$ is an absolutely continuous random variable with density $f(x) = \frac{1}{3}\mathbb{I}_{(0,3]}$. Then,

$$E(X) = \frac{2}{5}E(Y) + \frac{3}{5}E(Z) = \frac{2}{5} \times 0 + \frac{3}{5} \times \frac{3}{2} = \frac{9}{10}.$$

**9.49** Suppose that the distribution $P_X$ of $X$ is a mixture of a discrete probability distribution $Q$ and an absolutely continuous probability distribution $R$,

$$P_X = aQ + bR,$$

for some $a, b \geqslant 0$ such that $a + b = 1$. Since $Q$ is a discrete distribution, there is a countable set $A \subset \mathbb{R}$ such that $Q(A) = 1$. Since $R$ is an absolutely continuous distribution, $R(B) = 0$ for any Borel set $B \subset \mathbb{R}$ of Lebesgue measure zero.

Let $Y_1, Y_2, \ldots$ be the results of consecutive coin tosses: 1 for heads and 0 for tails. Then,

$$X = \frac{Y_1}{3} + \frac{Y_2}{3^2} + \frac{Y_3}{3^3} + \cdots .$$

It follows that all values of $X$ belong to the set

$$C = \left\{ \tfrac{y_1}{3} + \tfrac{y_2}{3^2} + \tfrac{y_3}{3^3} + \cdots : y_k = 0 \text{ or } 1 \text{ for each } k = 1, 2, \ldots \right\};$$

that is, $P_X(C) = 1$. It is not hard to verify that $C$ is the Cantor set of Problem 5.23. We know by Problem 6.20 that $C$ is of Lebesgue measure zero, so $R(C) = 0$. As a result, $X$ must be a discrete random variable with $P_X = Q$.

However, for any $x \in \mathbb{R}$, we have $P_X(\{x\}) = 0$. Indeed, if $x \notin C$, then $\{X = x\} = \emptyset$ and $P(\{X = x\}) = 0$. If $x \in C$, then $x = \frac{y_1}{3} + \frac{y_2}{3^2} + \frac{y_3}{3^3} + \cdots$ for some $y_1, y_2, \ldots \in \{0, 1\}$ and

$$\begin{aligned} P(\{X = x\}) &= P(\{Y_1 = y_1, Y_2 = y_2, \ldots\}) \\ &= \lim_{n\to\infty} P(\{Y_1 = y_1, \ldots, Y_n = y_n\}) = \lim_{n\to\infty} \frac{1}{2^n} = 0. \end{aligned}$$

It follows that $P_X(A) = 0$ for any countable set $A \subset \mathbb{R}$. This means that $X$ cannot be a discrete random variable, leading to a contradiction. As a result, the distribution of $X$ is neither discrete nor absolutely continuous nor a mixture thereof.

**9.50** Let $X_n$ be the total gain after $n$ tosses of the coin and let $Y_k$ be the gain at the $k$th toss of the coin, so $X_n = Y_1 + \cdots + Y_n$. Clearly,

$E(Y_k) = \frac{1}{3^k}$ and

$$E(X_n) = E(Y_1) + \cdots + E(Y_n)$$
$$= \frac{1}{3} + \frac{1}{3^2} + \cdots + \frac{1}{3^n} = \frac{1}{2}\left(1 - \frac{1}{3^n}\right).$$

Since $X_n \nearrow X$, it follows that

$$E(X) = \lim_{n\to\infty} E(X_n) = \lim_{n\to\infty} \frac{1}{2}\left(1 - \frac{1}{3^n}\right) = \frac{1}{2}.$$

# 10

# Conditional Expectation

## 10.1 Theory and Problems

The general definition of conditional expectation will be preceded by a series of examples designed to develop the construction step by step.

As we have learned in Chapter 7, the knowledge of whether or not a certain event has occurred may influence the odds of other events. As a result, it will have an effect on the expectation of random variables.

**10.1** A fair die with faces $1, 2, 3$ colored green and faces $4, 5, 6$ colored red is rolled once. Let $X$ be the number shown.

a) What is the expectation of $X$?

b) If you can see that the die has landed green face up (but cannot see the actual number shown), what will the expected value of $X$ be in these circumstances?

The solution to Problem 10.1 suggests the following definition.

▶ **Definition 10.1** If $X$ is a discrete random variable with different values $x_1, x_2, \ldots$ and $B$ is an event such that $P(B) \neq 0$, then the *conditional expectation* of $X$ given $B$ is defined by

$$E(X|B) = \sum_{i=1}^{\infty} x_i P(\{X = x_i\}|B),$$

provided that the series is absolutely convergent.

**10.2** Show that $E(\mathbb{I}_A|B) = P(A|B)$ if $P(B) \neq 0$. (As usual, $\mathbb{I}_A$ denotes the indicator function of an event $A$.)

**10.3** A fair die is rolled twice. What is the conditional expectation of the sum of outcomes given the first roll shows 1?

**10.4** We toss a coin repeatedly until heads appear, hitting the numerical keypad of a calculator randomly after each toss. Find the conditional expectation of the number keyed in, given that heads appear for the first time after $i$ tosses. (Our calculator is a special model capable of displaying infinitely many digits.)

**10.5** Let $X$ be a random variable with the Poisson distribution. Find the conditional expectation of $X$ given that $X$ is an even number.

**10.6** For any discrete random variable $X$ and any event $B$ such that $P(B) \neq 0$, show that

$$E(X|B) = \frac{E(\mathbb{I}_B X)}{P(B)}, \tag{10.1}$$

where $\mathbb{I}_B$ is the indicator function of $B$.

Formula (10.1) can often help in computing the conditional expectation of discrete random variables. In the problems below, find the conditional expectation $E(X|B_i)$.

**10.7** $X$ is the sum of outcomes obtained by rolling a die twice and $B_i$ is the event that the first die shows $i$.

**10.8** $X$ denotes the amount shown if three coins, 5¢, 10¢, and 25¢, are tossed, and $B_i$ denotes the event that $i$ coins land heads up.

∘ **10.9** $X$ denotes the number of points scored by a chess player in a match against an equally strong opponent, where draws are discarded so that the odds of winning or losing a game are fifty-fifty, the match ends if one of the players reaches 3 points, and $B_i$ is the event that $i$ games have been played.

Formula (10.1) can also be used to extend the definition of expectation conditioned on an event to the case of an arbitrary random variable.

▸ **Definition 10.2** For any random variable $X$ and any event $B$ such that $P(B) \neq 0$, the *conditional expectation* of $X$ given $B$ is defined by

$$E(X|B) = \frac{E(\mathbb{I}_B X)}{P(B)},$$

provided that the expectation on the right-hand side exists.

**10.10** Let $X$ be a random variable with exponential distribution. Find $E(X|\{X \geqslant t\})$.

**10.11** Let $X$ and $Y$ be jointly continuous random variables with joint density

$$f_{X,Y}(x,y) = \begin{cases} e^{-x-y} & \text{if } x, y \geqslant 0 \\ 0 & \text{otherwise.} \end{cases}$$

Compute the conditional expectation of $X + Y$ given that $X < Y$.

* **10.12** Show that if $E(X)$ is exists, then so does $E(X|B)$ for any event $B$ such that $P(B) \neq 0$.

In many situations, we can observe the value of a certain random variable $Y$. Once the value of $Y$ becomes known, it may have an effect on the expectation of another random variable $X$. We shall be working toward a reasonable definition of the conditional expectation of $X$ given $Y$.

In particular, if $Y$ is a discrete random variable, then one of its values $y_1, y_2, \ldots$ may be observed. Once it becomes known that $Y = y_i$, say, then we would take the conditional expectation of $X$ given the event $\{Y = y_i\}$. However, we may not know the value of $Y$ in advance. In such circumstances, the best one can do is to consider all possibilities, i.e., to make a list of possible conditional expectations:

$$E(X|\{Y = y_1\}), E(X|\{Y = y_1\}), \ldots .$$

A convenient way of doing this is to construct a new discrete random variable constant on $\{Y = y_i\}$ and equal to $E(X|\{Y = y_i\})$ for each $i$. This idea is captured in the next definition.

▶ **Definition 10.3** Let $X$ be an arbitrary random variable and let $Y$ be a discrete random variable with different values $y_1, y_2, \ldots$ . Then, the *conditional expectation* $E(X|Y)$ of $X$ given $Y$ is defined to be a random variable such that for any $i = 1, 2, \ldots,$

$$E(X|Y)(\omega) = E(X|\{Y = y_i\}) \quad \text{for every } \omega \in \{Y = y_i\}. \tag{10.2}$$

Clearly, $E(X|Y)$ defined in this way is a discrete random variable with values $E(X|\{Y = y_1\}), E(X|\{Y = y_1\}), \ldots$ .

Find an explicit formula for $E(X|Y)$ in the problems below. How may different values does $E(X|Y)$ take?

**10.13** A die is rolled twice; $X$ is the sum of the outcomes and $Y$ is the outcome of the first roll.

**10.14** Three coins, 5¢, 10¢, and 25¢, are tossed; $X$ is the total amount shown and $Y$ is the number of heads.

**10.15** From a bag containing four balls numbered 1, 2, 3, and 4 you draw two. If at least one of the numbers drawn is greater than 2, you win \$1, and otherwise you lose \$1. Let $X$ be the amount won or lost and let $Y$ be the first number drawn.

**10.16** Let $\Omega = \{1, 2, \dots\}$ be equipped with the $\sigma$-field of all subsets of $\Omega$ and a probability measure $P$ such that $P(\{k\}) = 2^{-k}$ for each $k = 1, 2, \dots$ . Find $E(X|Y)$ if $X(k) = k$ and $Y(k) = (-1)^k$ for each $k = 1, 2, \dots$ .

∘ **10.17** A coin is tossed three times behind a curtain and you win an amount $X$ equal to the number of heads shown. If, peeping behind the curtain, you could see the result $Y$ of the first toss, how would you assess your expected winnings?

∘ **10.18** A bag contains three coins, but one has heads on both sides. Two coins are drawn and tossed. Find $E(X|Y)$ if $X$ is the number of heads and $Y$ is the number of genuine coins among the drawn ones.

**10.19** Show that if $X$ and $Y$ are independent random variables and $Y$ is discrete, then $E(X|Y) = E(X)$.

We have learned in Chapter 8 that whether a function $Y : \Omega \to \mathbb{R}$ is a random variable or not depends on the choice of the $\sigma$-field on $\Omega$. The smallest $\sigma$-field with respect to which $Y$ is a random variable plays a special role in the construction of conditional expectation. It will allow us to discard the restriction in Definition 10.3 that the conditioning random variable $Y$ should be discrete. As in Chapter 8, we denote by $\{Y \in B\}$ the inverse image of a set $B \subset \mathbb{R}$ under $Y$.

▶ **Definition 10.4** Let $Y$ be a random variable. The family consisting of all inverse images $\{Y \in B\}$, where $B$ is a Borel set in $\mathbb{R}$, is denoted by $\mathcal{F}_Y$ and called the *$\sigma$-field generated by* $Y$. (This family is a $\sigma$-field by Problem 5.8.)

A convenient way to think of $\mathcal{F}_Y$ is that it consists of all events about which it will be possible to tell whether they have occurred or not once the value of $Y$ becomes known.

**10.20** Show that $\mathcal{F}_Y$ is the smallest $\sigma$-field with respect to which $Y$ is a random variable.

Find and compare the $\sigma$-fields generated by the random variables $Y$ and $E(X|Y)$ in the problems below. How may elements do these $\sigma$-fields have?

**10.21** A die is rolled twice; $Y$ is the first outcome and $X$ is the sum of outcomes.

**10.22** Three coins, 5¢, 10¢, and 25¢, are tossed and you can take those showing heads; $X$ is the total amount you take and $Y$ is the number of heads.

**10.23** From a bag containing four balls numbered 1, 2, 3, and 4, you draw two; if at least one of the numbers drawn is greater than 2, you win \$1, and otherwise you lose \$1. Let $X$ be the amount won or lost and let $Y$ be the first number drawn.

∘ **10.24** A die is rolled twice; if the sum of outcomes is even, you win \$1, and if it is odd, you lose \$1; $X$ is the amount won or lost and $Y$ is the outcome of the first roll.

∘ **10.25** A die is rolled twice; $Y$ is the outcome of the first roll and $X$ is the outcome of the second roll.

In the problems above, the $\sigma$-field $\mathcal{F}_{E(X|Y)}$ turns out to be contained in $\mathcal{F}_Y$. This is connected with the fact that the random variable $E(X|Y)$ is never "more complicated" than $Y$ in the sense that whenever $Y$ is constant on a set, so is $E(X|Y)$. This observation leads to the following general property.

**10.26** Let $Y$ be a discrete random variable. Show that for $E(X|Y)$ given by Definition 10.3,

$$\mathcal{F}_{E(X|Y)} \subset \mathcal{F}_Y.$$

The value of $E(X|Y)$ on each set of the form $\{Y = i\}$ can be thought of as the mean value of $X$ over this set. Let us take the mean of all such mean values. The next problem shows that this "mean of means" is simply $E(X)$ !

**10.27** Let $Y$ be a discrete random variable. Show that

$$E(E(X|Y)) = E(X).$$

It turns out that a much stronger property holds; namely $E(X|Y)$ is close to $X$ in the sense that their means over certain events smaller than the whole probability space are also equal. Here, by the mean of $X$ over an event $B$, we understand the expectation $E(\mathbb{I}_B X)$, and similarly for the mean of $E(X|Y)$ over $B$.

**10.28** Let $Y$ be a discrete random variable. For any $B \in \mathcal{F}_Y$, prove that

$$E(\mathbb{I}_B E(X|Y)) = E(\mathbb{I}_B X).$$

We are now well prepared for the main definition of this chapter. The price we have to pay for relaxing the assumption that the conditioning random variable $Y$ should be discrete is that we shall no longer have an explicit formula for $E(X|Y)$ as in Definition 10.3. Instead, the conditional expectation $E(X|Y)$ will be defined implicitly by imposing the conditions in Problems 10.26 and 10.28 in the case of an arbitrary $Y$.

▶ **Definition 10.5** Let $X$ and $Y$ be any random variables. The *conditional expectation* $E(X|Y)$ of $X$ given $Y$ is a random variable such that

(a) $\mathcal{F}_{E(X|Y)} \subset \mathcal{F}_Y$,

(b) $E(\mathbb{I}_B E(X|Y)) = E(\mathbb{I}_B X)$ for any $B \in \mathcal{F}_Y$.

By Problems 10.26 and 10.28, this definition is consistent with Definition 10.3.

Conditions (a) and (b) impose conflicting restrictions on $E(X|Y)$. On the one hand, $E(X|Y)$ needs to have a rich enough structure to satisfy (b). On the other hand, it cannot be too rich or the $\sigma$-field $\mathcal{F}_{E(X|Y)}$ would be too large to satisfy (a). Meeting the two conditions simultaneously calls for a compromise. That such a compromise can always be found is a deep result in measure theory called the Radon-Nikodym theorem. Moreover, the conditional expectation $E(X|Y)$ is unique to within a modification on a set of probability zero. This means that we can alter the values of $E(X|Y)$ on a set of probability zero and the resulting new random variable will also be a valid conditional expectation, but this is no longer the case if the values of $E(X|Y)$ are altered on a set of positive probability.

▶ **Theorem 10.1** *If $X$ is a random variable such that $E(|X|) < \infty$ and $Y$ is an arbitrary random variable, then the conditional expectation $E(X|Y)$ exists and is unique to within a modification on a set of probability zero.*

The proof of this theorem is beyond the scope of this book. Here, we shall restrict ourselves to finding the conditional expectation in some concrete cases.

In the problems below, take $\Omega = [0, 1]$ with the $\sigma$-field $\mathcal{B}$ of Borel sets and $P$ the Lebesgue measure on $[0, 1]$, and find $E(X|Y)$.

**10.29** $X(\omega) = 2\omega$, $Y(\omega) = 2$.

**10.30** $X(\omega) = 2\omega$ for all $\omega \in [0, 1]$; $Y(\omega) = 0$ if $\omega \in [0, \frac{1}{2})$ and $Y(\omega) = 1$ if $\omega \in [\frac{1}{2}, 1]$.

**10.31** $X(\omega) = 2\omega$ and $Y(\omega) = \omega^2$.

**10.32** $X(\omega) = 2\omega$; $Y(\omega) = \omega$ if $\omega \in [0, \frac{1}{2})$ and $Y(\omega) = \frac{1}{2}$ if $\omega \in [\frac{1}{2}, 1]$.

**10.33** $X = 2\omega - 1 + |2\omega - 1|$ and $Y = 1 - |2\omega - 1|$.

∘ **10.34** $X(\omega) = 1 - \omega$; $Y(\omega) = 0$ if $\omega \in [0, \frac{1}{2})$ and $Y(\omega) = 1$ if $\omega \in [\frac{1}{2}, 1]$.

∘ **10.35** $X(\omega) = 1 - \omega$; $Y(\omega) = 0$ if $\omega \in [0, \frac{1}{2})$ and $Y(\omega) = \omega$ if $\omega \in [\frac{1}{2}, 1]$.

The following property extends that in Problem 10.27 to arbitrary random variables.

▶ **10.36** Show that for any $X$ and $Y$,

$$E(E(X|Y)) = E(X).$$

The last property can be refined as follows. The formula below is known as the *tower* property or the *iterated conditional expectation* property.

▶ **10.37** Prove that for any random variables $X$, $Y$, and $Z$ such that $\mathcal{F}_Y \subset \mathcal{F}_Z$,

$$E(E(X|Z)|Y) = E(X|Y).$$

Problem 10.19 can also be extended to the case of arbitrary random variables.

▶ **10.38** Prove that if $X$ and $Y$ are independent random variables, then

$$E(X|Y) = E(X).$$

Here is yet another important property of conditional expectations. It generalizes the formula $E(aX) = aE(X)$, where $a$ is a number. In the case of conditional expectation, we can replace $a$ by any random variable $Z$ as long as $\mathcal{F}_Z \subset \mathcal{F}_Y$, where $Y$ is the conditioning random variable. The inclusion means that whenever the value of $Y$ is known, the value of $Z$ is known too. Because of this, the formula below of often referred to as "*taking out what is known.*"

▶ **10.39** Assuming that $\mathcal{F}_Z \subset \mathcal{F}_Y$, prove that

$$E(ZX|Y) = ZE(X|Y).$$

We turn our attention to jointly continuous random variables $X$ and $Y$. In this case, the conditional expectation $E(X|Y)$ can be expressed in terms of the joint density $f_{X,Y}$.

▶ **Definition 10.6** We put

$$h(x|y) = \frac{f_{X,Y}(x,y)}{f_Y(y)}$$

if $f_Y(y) \neq 0$, and $h(x|y) = 0$ otherwise, and we call $h(x|y)$ the *conditional density* of $X$ given $Y$.

By $h(x|Y)$ we shall denote the random variable assigning $h(x|Y(\omega))$ to each $\omega \in \Omega$.

▶ **10.40** Prove that

$$E(X|Y) = \int_{-\infty}^{+\infty} x h(x|Y)\, dx. \tag{10.3}$$

**10.41** Show without referring to Problem 10.38 that $E(X|Y) = E(X)$ if $X$ and $Y$ are jointly continuous and independent.

Find $E(X|Y)$ if the joint density of $X$ and $Y$ is of the following form with a constant $K$ such that $\int_{-\infty}^{+\infty}\int_{-\infty}^{+\infty} f_{X,Y}(x,y)\, dx\, dy = 1$.

**10.42** $f_{X,Y}(x,y) = K(x+y)$ for $0 \leqslant x, y \leqslant 1$ and zero otherwise.

**10.43** $f_{X,Y}(x,y) = K \cos x \cos y$ for $0 \leqslant x, y \leqslant \frac{\pi}{2}$ and zero otherwise.

**10.44** $f_{X,Y}(x,y) = K(2x+y)$ for $x^2 + y^2 \leqslant 1$ and zero otherwise.

∘ **10.45** $f_{X,Y}(x,y) = K(x+y)^2$ for $0 \leqslant x, y \leqslant 1$ and zero otherwise.

∘ **10.46** $f_{X,Y}(x,y) = K(x+y^2)$ for $x^2 + y^2 \leqslant 1$ and zero otherwise.

The remainder of this chapter is devoted to various problems involving conditional expectation.

**10.47** In a game played on a 3-square board a coin is tossed repeatedly: heads allow the player to move to the next square and tails to jump over the next square. If the end of the board is reached, the reward is \$10. However, there is a trap on one of the squares, placed randomly with uniform probability. If the player falls into the trap, the game stops and he has to pay a penalty of \$10. Let $X$ be the gain or loss made by the player and let $Y$ be the number of the square containing the trap. Compute $E(X|Y)$ and $E(X)$.

**10.48** We buy $Y$ raffle tickets, where $Y$ has the Poisson distribution with parameter $\lambda$. Each ticket wins a prize with probability $p$ independently of the others. Find the conditional expectation of the number of prizes won $X$ given the number of tickets bought $Y$, and compute the expectation of $X$.

**10.49** At a given moment, there are $N$ trucks on a bridge. Let $Y_i$ be the weight of the $i$th truck and let $X$ be the total weight of the trucks. Show that if $N$ and all the $Y_i$ are independent, $E(Y_i) = m$, and $E(N) = n$, then $E(X) = mn$.

**10.50** The number of mistakes made by a typesetter on any given page is a random variable having the Poisson distribution with parameter $\lambda$, independent for each page. A three-page article is prepared by one of four typesetters, for whom the values of $\lambda$ are 1, 2, 3, and 4, respectively. The typesetter is selected randomly with uniform probability. What is the expected number of misprints throughout the article?

∘ **10.51** Suppose that the number of customers in a shop on a given day has the Poisson distribution with parameter $\lambda$. Each customer makes a purchase with probability $p$ independently of other customers. What is the expected number of customers who will make a purchase today?

∘ **10.52** The number of customers visiting a shop on a given day has the Poisson distribution with parameter $\lambda$. The expected number of items bought by each customer is 2, the purchases of each customer being independent. What is the expected number of items the shop will sell in a day?

## 10.2 Hints

**10.1** a) See Problem 9.1.
b) When computing the expectation, instead of the probability of any given outcome, use the conditional probability of that outcome given that the die has landed green face up.

**10.2** What are the values of $\mathbb{I}_A$? What are the corresponding conditional probabilities given $B$?

**10.3** List all the values of the sum of outcomes. What are the corresponding conditional probabilities given $B$?

**10.4** What is the range of possible numbers keyed in, given that heads appear after $i$ tosses? What is the conditional probability of obtaining any one of these numbers, given that heads appear for the first time after $i$ tosses?

**10.5** You may need the Taylor expansions of $\cosh x$ and $\sinh x$ about $x = 0$.

**10.6** Each value of $\mathbb{I}_B X$ is equal to one of the values of $X$ or to 0. Observe that if $x$ is a non-zero value of $\mathbb{I}_B X$, then

$$\frac{P(\{\mathbb{I}_B X = x\})}{P(B)} = \frac{P(\{X = x\} \cap B)}{P(B)} = P(\{X = x\}|B).$$

**10.7** What are the values of $X$ on $B_i$? What are the corresponding probabilities? Now, find $E(\mathbb{I}_{B_i}X)$, $P(B_i)$, and, finally, $E(X|B_i)$.

**10.8** See Problem 10.7.

**10.9** See Problem 10.7.

**10.10** Consider the cases when $t < 0$ and $t \geqslant 0$ separately.

**10.11** Express both $P(\{X < Y\})$ and $E(\mathbb{I}_{\{X<Y\}}(X+Y))$ in terms of the joint density $f_{X,Y}(x,y)$ as integrals over the half-plane defined by the inequality $x < y$.

**10.12** You need to show that $E(\mathbb{I}_B X)$ exists if $E(X)$ does. Consider the negative and positive parts of $\mathbb{I}_B X$.

**10.13** What are the values of $Y$? Refer to Problem 10.7 to find the corresponding values of $E(X|Y)$.

**10.14** List all the values of $Y$. Refer to Problem 10.8 to find the corresponding values of $E(X|Y)$.

**10.15** See Problems 10.13 and 10.14.

**10.16** How many values does the random variable $Y$ take? What are the corresponding values of $E(X|Y)$?

**10.17** Peeping behind the curtain, you could find out the value of $Y$ (1 if heads and 0 if tails, say). Therefore, the conditional expectation you are looking for will be a random variable with at most two values; cf. Problems 10.13, 10.14, and 10.15.

**10.18** The random variable $Y$ can take two different values 1 or 2. The random variable $E(X|Y)$ will take at most two different values; cf. Problems 10.13, 10.14, and 10.15.

**10.19** If $X$ and $Z$ are independent random variables, then $E(XZ) = E(X)E(Z)$. Use $Z = \mathbb{I}_{\{Y=y\}}$.

**10.20** You need to show that $\mathcal{F}_Y$ is contained in every $\sigma$-field $\mathcal{G}$ with respect to which $Y$ is a random variable.

**10.21** The inverse images of one-element sets will give you the atoms of the $\sigma$-fields generated by $Y$ and, respectively, by $E(X|Y)$. See Problems 3.21–3.26 to recall the properties of atoms.

**10.22** The values of $Y$ and $E(X|Y)$ are listed in Solution 10.14. The inverse images of these values are the atoms of the $\sigma$-fields generated by $Y$ and $E(X|Y)$, respectively.

**10.23** Here, $E(X|Y)$ has fewer values than $Y$; see Problem 10.15. As a result, the $\sigma$-field generated by $E(X|Y)$ will be smaller than that generated by $Y$.

**10.24** See Problems 10.21, 10.22, and 10.23.

**10.25** See Problems 10.21, 10.22, and 10.23.

**10.26** Show that $\mathcal{F}_Y$ is generated by countably many atoms. Observe that $E(X|Y)$ is constant on each of these atoms.

**10.27** Let $y_1, y_2 \ldots$ be the values of $Y$. Take the value of $E(X|Y)$ on each set $\{Y = y_i\}$, multiply by the probability of this set, and add up for all $i$. Use monotone convergence (condition (c) in Definition 9.7) to show that the resulting sum is equal to $E(X)$.

**10.28** First, consider $B$ to be an event of the form $\{Y = y_i\}$, where $y_i$ is one of the values of $Y$. Then, take $B$ equal to a finite union of such events. Finally, use monotone convergence (condition (c) in Definition 9.7) to prove the equality for any $B$ being a countable union of events of the form $\{Y = y_i\}$. Observe that such countable unions exhaust the $\sigma$-field $\mathcal{F}_Y$.

**10.29** The random variable $Y$ is constant; hence, $\mathcal{F}_Y = \{\emptyset, \Omega\}$, which implies a restriction on $E(X|Y)$.

**10.30** The random variable $Y$ has two different values, so $E(X|Y)$ can have at most two different values.

**10.31** $\mathcal{F}_Y$ is large in this case and condition (a) poses no restriction on $E(X|Y)$.

**10.32** The conditional expectation $E(X|Y)$ on $[\frac{1}{2}, 1]$ can be found in a similar way as in Solution 10.30 because $Y$ is constant on $[\frac{1}{2}, 1]$. To find $E(X|Y)$ on $[0, \frac{1}{2})$, use a similar argument as in Solution 10.31.

**10.33** Observe that the graph of $Y$ is symmetric about 1/2. Does it mean that all events in $\mathcal{F}_Y$ must also be symmetric about 1/2? If so, does it mean that the graph of $E(X|Y)$ must be symmetric about 1/2?

**10.34** See Problem 10.30.

**10.35** See Problem 10.32.

**10.36** Use condition (b) of Definition 10.5 with $B = \Omega$.

**10.37** You need to demonstrate that conditions (a) and (b) in Definition 10.5 remain satisfied if every occurrence of $E(X|Y)$ is replaced by $E(E(X|Z)|Y)$; that is to say, you need to verify that:
(a) $\mathcal{F}_{E(E(X|Z)|Y)} \subset \mathcal{F}_Y$,
(b) $E(\mathbb{I}_B E(E(X|Z)|Y)) = E(\mathbb{I}_B X)$ for any $B \in \mathcal{F}_Y$.
To deduce that $E(E(X|Z)|Y) = E(X|Y)$ you may use the uniqueness of conditional expectation asserted in Theorem 10.1.

**10.38** Show that conditions (a) and (b) in Definition 10.5 remain satisfied if every occurrence of $E(X|Y)$ is replaced by $E(X)$.

**10.39** First, verify the formula for $Z = \mathbb{I}_C$, then extend it to any random variable $Z$ with finitely many values, and, finally, to an arbitrary $Z$ using monotone convergence (condition (c) in Definition 9.7).

**10.40** Verify the conditions of Definition 10.5. You may need Theorem 9.1 to express certain expectations in terms of the density.

**10.41** What does the independence of $X$ and $Y$ mean in terms of their joint density?

**10.42** Find the density $f_Y(y)$ of $Y$ and then compute the integral in (10.3), bearing in mind that the limits of integration are finite, since in this case, $f_{X,Y}(x, y)$ is zero outside a rectangle.

**10.43** Use formula (10.3).

**10.44** Use formula (10.3).

**10.45** See Problems 10.42–10.44.

**10.46** See Problems 10.42–10.44.

**10.47** How may values does $Y$ have? Find the corresponding values of $E(X|Y)$. Then, the formula in Problem 10.36 can be used to compute $E(X)$.

**10.48** List all the values of the discrete random variable $Y$ and find the corresponding values of $E(X|Y)$. Then, compute $E(X)$ using the formula in Problem 10.36.

**10.49** This is similar to Problem 10.48. The main difference is that the distribution of $N$ is unknown, whereas previously we were dealing with a specific (Poisson) distribution.

**10.50** Suppose that the typesetter with $Y = \lambda$, where $\lambda = 1, 2, 3, 4$, is selected. Find the conditional expectation of the number of misprints on a single page given $Y$, then find the conditional expectation of the total number of misprints in the article given $Y$, and, finally, apply the formula in Problem 10.36 to compute the expected number of misprints throughout the article.

**10.51** See Problem 10.48.

**10.52** See Problem 10.49.

## 10.3 Solutions

**10.1** a) The random variable $X$ can take six values, $1, 2, 3, 4, 5, 6$, with probability $\frac{1}{6}$ each. As a result,

$$\begin{aligned}E(X) &= \sum_{k=1}^{6} kP(\{X = k\}) \\ &= 1 \times \frac{1}{6} + 2 \times \frac{1}{6} + 3 \times \frac{1}{6} + 4 \times \frac{1}{6} + 5 \times \frac{1}{6} + 6 \times \frac{1}{6} = 3.5\ .\end{aligned}$$

b) Let $B$ denote the event that the die has landed green face up. If we know that $B$ has occurred, then the probabilities $P(\{X = k\})$ in a) will need to be replaced by the conditional probabilities

$$\tilde{P}(\{X = k\}) = P(\{X = k\}|B).$$

The conditional probabilities of the outcomes 1, 2, 3, 4, 5, and 6 are respectively $\frac{1}{3}$, $\frac{1}{3}$, $\frac{1}{3}$, 0, 0, and 0. This gives a new value for the expectation

$$\begin{aligned}\tilde{E}(X) &= \sum_{k=1}^{6} k\tilde{P}(\{X = k\}) = \sum_{k=1}^{6} kP(\{X = k\}|B) \\ &= 1 \times \frac{1}{3} + 2 \times \frac{1}{3} + 3 \times \frac{1}{3} + 4 \times 0 + 5 \times 0 + 6 \times 0 = 2\ .\end{aligned}$$

**10.2** The indicator function $\mathbb{I}_A$ takes just two values 0 and 1. It follows that

$$\begin{aligned}E(\mathbb{I}_A|B) &= 0 \times P(\{\mathbb{I}_A = 0\}|B) + 1 \times P(\{\mathbb{I}_A = 1\}|B) \\ &= P(\{\mathbb{I}_A = 1\}|B) = P(A|B).\end{aligned}$$

**10.3** Let $B$ be the event that the first roll shows 1. Clearly, $P(B) = \frac{1}{6}$. The sum $X$ of outcomes can take 11 different values, $2, 3, \ldots, 12$. However, if we know that $B$ has occurred, then the sum cannot be greater than 7. This means that

$$P(\{X = i\}|B) = 0 \quad \text{for } i = 8, \ldots, 12.$$

On the other hand,

$$P(\{X = i\}|B) = \frac{P(\{X = i\} \cap B)}{P(B)} = \frac{1/36}{1/6} = \frac{1}{6} \quad \text{for } i = 2, \ldots, 7,$$

since $P(\{X = i\} \cap B)$ is the probability that the outcome of first roll is 1 and the outcome of the second roll is $i - 1$. It follows that

$$E(X|B) = \sum_{k=2}^{12} kP(\{X = k\}|B) = \sum_{k=2}^{7} \frac{k}{6} = 4.5\ .$$

**10.4** Let $X$ be the number keyed in and let $B_i$ denote the event that heads appear for the first time after $i$ tosses. On $B_i$, the random variable $X$ can take any one of the values $0, 1, 2, \ldots, 10^i - 1$, and for each such value $k$, we have $P(\{X = k\}|B_i) = 10^{-i}$. It follows that

$$E(X|B_i) = \sum_{k=0}^{10^i-1} kP(\{X = k\}|B_i) = \sum_{k=0}^{10^i-1} \frac{k}{10^i} = \frac{(10^i - 1)}{2}.$$

**10.5** Let $X$ have the Poisson distribution with parameter $\lambda$. Then,

$$P(\{X \text{ is even}\}) = \sum_{k=0}^{\infty} P(\{X = 2k\}) = \sum_{k=0}^{\infty} \frac{\lambda^{2k}}{(2k)!} e^{-\lambda} = e^{-\lambda} \cosh \lambda$$

and

$$P(\{X = 2k\}|\{X \text{ is even}\}) = \frac{P(\{X = 2k\})}{P(\{X \text{ is even}\})} = \frac{\lambda^{2k}}{(2k)! \cosh \lambda}.$$

It follows that

$$\begin{aligned} E(X|\{X \text{ is even}\}) &= \sum_{k=0}^{\infty} 2k \frac{\lambda^{2k}}{(2k)! \cosh \lambda} \\ &= \frac{\lambda}{\cosh \lambda} \sum_{k=0}^{\infty} \frac{\lambda^{2k-1}}{(2k-1)!} = \frac{\lambda \sinh \lambda}{\cosh \lambda}. \end{aligned}$$

**10.6** Suppose that $X$ has different values $x_1, x, \ldots$ . Then,

$$E(X|B) = \sum_{i=1}^{\infty} x_i P(\{X = x_i\}|B) = \frac{1}{P(B)} \sum_{i=1}^{\infty} x_i P(\{X = x_i\} \cap B).$$

Let us put

$$I = \{i : x_i \neq 0 \text{ and } \{X = x_i\} \cap B \text{ is non-empty}\}.$$

If $i \in I$, then $x_i P(\{X = x_i\} \cap B) = x_i P(\{\mathbb{I}_B X = x_i\})$. If $i \notin I$, then $x_i P(\{X = x_i\} \cap B) = 0$. It follows that

$$E(X|B) = \frac{1}{P(B)} \sum_{i \in I} x_i P(\{\mathbb{I}_B X = x_i\}) = \frac{E(\mathbb{I}_B X)}{P(B)}.$$

The last equality holds because all the non-zero values of $\mathbb{I}_B X$ are of the form $x_i$, where $i \in I$.

**10.7** In the case $i = 1$, see Solution 10.3. We shall use formula (10.1) to find a solution for any $i = 1, \ldots, 6$. If the first die shows $i$, then the sum must have one of the values $i+1, \ldots, i+6$, which are the non-zero values of $\mathbb{I}_{B_i}X$. Clearly, $\mathbb{I}_{B_i}X = i + j$ if and only if the outcome of the first roll is $i$ and that of the second roll is $j$, the probability of which is $\frac{1}{36}$. It follows that

$$E(\mathbb{I}_{B_i}X) = \sum_{j=1}^{6}(i+j)P(\{\mathbb{I}_{B_i}X = i+j\}) = \sum_{j=1}^{6}\frac{i+j}{36} = \frac{7}{12} + \frac{1}{6}i.$$

Because $P(B_i) = \frac{1}{6}$, we find that

$$E(X|B_i) = \frac{E(\mathbb{I}_{B_i}X)}{P(B_i)} = \frac{7/12 + i/6}{1/6} = 3.5 + i.$$

**10.8** First, we find $P(B_0) = P(B_3) = \frac{1}{8}$ and $P(B_1) = P(B_2) = \frac{3}{8}$. Next,

$$E(\mathbb{I}_{B_0}X) = 0$$

because $X$ is 0 on $B_0$,

$$E(\mathbb{I}_{B_1}X) = 5 \times \frac{1}{8} + 10 \times \frac{1}{8} + 25 \times \frac{1}{8} = 5$$

because the values of $X$ on $B_1$ are 5, 10, and 25, each with probability $\frac{1}{8}$,

$$E(\mathbb{I}_{B_2}X) = 15 \times \frac{1}{8} + 30 \times \frac{1}{8} + 35 \times \frac{1}{8} = 10$$

because the values of $X$ on $B_2$ are $5 + 10 = 15$, $10 + 20 = 30$, and $10 + 25 = 35$, each with probability $\frac{1}{8}$, and

$$E(\mathbb{I}_{B_3}X) = 40 \times \frac{1}{8} = 5$$

because the only value of $X$ on $B_3$ is $5 + 10 + 25 = 40$, with probability $\frac{1}{8}$. It follows that

$$E(X|B_0) = \frac{0}{1/8} = 0, \qquad E(X|B_1) = \frac{5}{3/8} = \frac{40}{3},$$

$$E(X|B_2) = \frac{10}{3/8} = \frac{80}{3}, \qquad E(X|B_3) = \frac{5}{1/8} = 40.$$

**10.10** Let $X$ be an exponential random variable with parameter $\lambda$. Since $X$ is non-negative, $\{X \geqslant t\} = \Omega$ if $t < 0$, and then $E(X|\{X \geqslant t\}) = E(X) = \frac{1}{\lambda}$; see Problem 9.38. Now, take any $t \geqslant 0$. In this case,

$$P(\{X \geqslant t\}) = \int_{-\infty}^{+\infty} f_X(s)\, ds = \int_t^{\infty} \lambda e^{-\lambda s}\, ds = e^{-\lambda t},$$

where

$$f_X(t) = \begin{cases} \lambda e^{-\lambda t} & \text{for } t \geqslant 0 \\ 0 & \text{for } t < 0, \end{cases}$$

is the density of $X$. Moreover,

$$\begin{aligned} E(\mathbb{I}_{\{X \geqslant t\}} X) &= \int_{-\infty}^{+\infty} \mathbb{I}_{\{s \geqslant t\}} s f_X(s)\, \mathrm{d}s = \int_t^{\infty} s\lambda e^{-\lambda s}\, \mathrm{d}s \\ &= te^{-\lambda t} + \frac{1}{\lambda} e^{-\lambda t}. \end{aligned}$$

It follows that

$$E(X|\{X \geqslant t\}) = \frac{E(\mathbb{I}_{\{X \geqslant t\}} X)}{E(\{X \geqslant t\})} = t + \frac{1}{\lambda}.$$

**10.11** First, we compute the probability

$$\begin{aligned} P(\{X < Y\}) &= \int_{-\infty}^{+\infty} \left( \int_x^{+\infty} f_{X,Y}(x,y)\, \mathrm{d}y \right) \mathrm{d}x \\ &= \int_0^{\infty} \left( \int_x^{\infty} e^{-x-y}\, \mathrm{d}y \right) \mathrm{d}x = \int_0^{\infty} e^{-2x} \mathrm{d}x = \frac{1}{2}. \end{aligned}$$

Next,

$$\begin{aligned} E(\mathbb{I}_{\{X<Y\}}(X+Y)) &= \int_{-\infty}^{+\infty} \left( \int_x^{+\infty} (x+y) f_{X,Y}(x,y)\, \mathrm{d}y \right) \mathrm{d}x \\ &= \int_0^{\infty} \left( \int_x^{\infty} (x+y) e^{-x-y}\, \mathrm{d}y \right) \mathrm{d}x \\ &= \int_0^{\infty} (2x+1)\, e^{-2x}\, \mathrm{d}x = 1. \end{aligned}$$

It follows that

$$E(X+Y|\{X<Y\}) = \frac{E(\mathbb{I}_{\{X<Y\}}(X+Y))}{P(\{X<Y\})} = \frac{1}{1/2} = 2.$$

**10.12** To show that $E(X|B) = E(\mathbb{I}_B X)/P(B)$ exists, we need to make sure that $E(\mathbb{I}_B X)$ does. This, in turn, requires that both $E(\mathbb{I}_B X^+)$ and $E(\mathbb{I}_B X^-)$ should exist. (For the definition of the positive and negative parts $X^+$ and $X^-$ of $X$, see Chapter 9.) Now, $E(\mathbb{I}_B X^+) \leqslant E(X^+)$ and $E(\mathbb{I}_B X^-) \leqslant E(X^-)$ because $\mathbb{I}_B X^+ \leqslant X^+$ and $\mathbb{I}_B X^- \leqslant X^-$. If $E(X)$ exists, then so do $E(X^+)$ and $E(X^-)$ and, consequently, $E(\mathbb{I}_B X^+)$ and $E(\mathbb{I}_B X^-)$ must exist, completing the proof.

**10.13** The random variable $Y$ can take six different values, $1, \ldots, 6$. In Solution 10.7, it was found that $E(X|\{Y=i\}) = 3.5 + i$. It follows that

$$E(X|Y)(\omega) = 3.5 + i$$

for every $\omega \in \{Y = i\}$, where $i = 1, \ldots, 6$; this is to say, the random variable $E(X|Y)$ has six different values.

**10.14** The random variable $Y$ can take four different values, $0, 1, 2, 3$. In Solution 10.8, it was found that

$$E(X|\{Y=0\}) = 0, \qquad E(X|\{Y=1\}) = \frac{40}{3},$$
$$E(X|\{Y=2\}) = \frac{80}{3}, \qquad E(X|\{Y=3\}) = 40.$$

It follows that $E(X|Y)$ has four different values, $0, 40/3, 80/3, 40$; namely

$$E(X|Y)(\omega) = \begin{cases} 0 & \text{if } \omega \in \{Y=0\} \\ 40/3 & \text{if } \omega \in \{Y=1\} \\ 80/3 & \text{if } \omega \in \{Y=2\} \\ 40 & \text{if } \omega \in \{Y=3\}. \end{cases}$$

**10.15** The random variable $Y$ has four different values, $1, 2, 3, 4$. The conditional expectations of $X$ given the first number drawn are

$$E(X|\{Y=1\}) = E(X|\{Y=2\}) = \frac{1}{3},$$
$$E(X|\{Y=3\}) = E(X|\{Y=4\}) = 1.$$

It follows that the random variable $E(X|Y)$ has just two values,

$$E(X|Y)(\omega) = \begin{cases} 1/3 & \text{if } \omega \in \{Y = 1 \text{ or } 2\} \\ 1 & \text{if } \omega \in \{Y = 3 \text{ or } 4\}. \end{cases}$$

**10.16** The random variable $Y$ takes only two values, $-1$ and $+1$, with probabilities, respectively,

$$P(\{Y=-1\}) = \sum_{k=1}^{\infty} P(\{Y = 2k-1\}) = \sum_{k=1}^{\infty} 2^{-2k+1} = \frac{2}{3},$$
$$P(\{Y=+1\}) = \sum_{k=1}^{\infty} P(\{Y = 2k\}) = \sum_{k=1}^{\infty} 2^{-2k} = \frac{1}{3}.$$

It follows that $E(X|Y)$ has at most two values,

$$E(X|\{Y=-1\}) = \frac{E(\mathbb{I}_{\{Y=-1\}}X)}{P(\{Y=-1\})} = \frac{1}{2/3}\sum_{k=1}^{\infty}(2k-1)2^{-2k+1} = \frac{5}{3},$$

$$E(X|\{Y=+1\}) = \frac{E(\mathbb{I}_{\{Y=+1\}}X)}{P(\{Y=+1\})} = \frac{1}{1/3}\sum_{k=1}^{\infty}2k2^{-2k} = \frac{8}{3},$$

and

$$E(X|Y)(k) = \begin{cases} 5/3 & \text{if } k \text{ is odd} \\ 8/3 & \text{if } k \text{ is even.} \end{cases}$$

**10.19** Let $y$ be one of the values of $Y$. It suffices to show that $E(X|Y)$ is equal to $E(X)$ on $\{Y=y\}$.

If $X$ and $Y$ are independent random variables, then so are $X$ and $\mathbb{I}_{\{Y=y\}}$. It follows that

$$E(\mathbb{I}_{\{Y=y\}}X) = E(\mathbb{I}_{\{Y=y\}})E(X) = P(\{Y=y\})E(X)$$

and

$$E(X|\{Y=y\}) = \frac{E(\mathbb{I}_{\{Y=y\}}X)}{P(\{Y=y\})} = \frac{P(\{Y=y\})E(X)}{P(\{Y=y\})} = E(X).$$

**10.20** We know that $\mathcal{F}_Y$ is a $\sigma$-field by Problem 5.8. Clearly, $Y$ is a random variable with respect to the $\sigma$-field $\mathcal{F}_Y$ because $\mathcal{F}_Y$ contains all sets of the form $\{Y \in B\}$, where $B$ is a Borel set in $\mathbb{R}$.

Now, if $Y$ is a random variable with respect to a $\sigma$-field $\mathcal{G}$, then $\{Y \in B\} \in \mathcal{G}$ for all Borel sets $B$. It follows that $\mathcal{F}_Y \subset \mathcal{G}$. This means that $\mathcal{F}_Y$ is the smallest $\sigma$-field with respect to which $Z$ is a random variable.

**10.21** The $\sigma$-field generated by $Y$ must contain the inverse images of all Borel sets and, in particular, the inverse images of all one-element sets in $\mathbb{R}$. The inverse images of one-element sets $\{a\}$ are empty unless $a$ is one of the values of $Y$. The six different values of $Y$ are $1, \ldots, 6$, the corresponding inverse images being $\{Y=1\}, \ldots, \{Y=6\}$. The $\sigma$-field $\mathcal{F}_Y$ is generated by these six atoms. Therefore, it has $2^6$ elements (see Problem 3.26).

We know from Solution 10.13 that $E(X|Y)$ has six values of the form $3.5+i$, where $i=1,\ldots,6$. It also follows from Solution 10.13 that the inverse images of these values are $\{E(X|Y)=3.5+i\} = \{Y=i\}$. This means that the $\sigma$-field $\mathcal{F}_{E(X|Y)}$ is generated by the same six atoms as $\mathcal{F}_Y$. Therefore, in this case, $\mathcal{F}_{E(X|Y)} = \mathcal{F}_Y$.

**10.22** The random variable $Y$ can take four different values, $0, 1, 2, 3$. The corresponding inverse images

$$\{Y = 0\},\ \{Y = 1\},\ \{Y = 2\},\ \{Y = 3\}$$

are the atoms generating the $\sigma$-field $\mathcal{F}_Y$. As a result, $\mathcal{F}_Y$ has $2^4$ elements (see Problem 3.26).

We know from Solution 10.14 that $E(X|Y)$ has four different values, $0, 40/3, 80/3, 40$, and that the corresponding inverse images are

$$\{E(X|Y) = 0\} = \{Y = 0\}, \qquad \{E(X|Y) = \tfrac{40}{3}\} = \{Y = 1\},$$
$$\{E(X|Y) = \tfrac{80}{3}\} = \{Y = 2\}, \qquad \{E(X|Y) = 40\} = \{Y = 3\},$$

so the $\sigma$-field $\mathcal{F}_{E(X|Y)}$ is generated by the same four atoms as $\mathcal{F}_Y$. As a result, $\mathcal{F}_{E(X|Y)} = \mathcal{F}_Y$.

**10.23** The random variable $Y$ has four different values, $1, 2, 3, 4$, so the four inverse images

$$\{Y = 1\},\ \{Y = 2\},\ \{Y = 3\},\ \{Y = 4\}$$

are the atoms generating the $\sigma$-field $\mathcal{F}_Y$ with $2^4 = 16$ elements.

From Solution 10.15 we know that $E(X|Y)$ has two different values, $1/3$ and $1$, the corresponding inverse images being

$$\{E(X|Y) = 1/3\} = \{Y = 1 \text{ or } 2\},$$
$$\{E(X|Y) = 1\} = \{Y = 3 \text{ or } 4\}.$$

It follows that the $\sigma$-field $\mathcal{F}_{E(X|Y)}$ is generated by two atoms

$$\{Y = 1 \text{ or } 2\},\ \{Y = 3 \text{ or } 4\}$$

and has $2^2 = 4$ elements. Moreover,

$$\mathcal{F}_{E(X|Y)} \subset \mathcal{F}_Y.$$

**10.26** If $Y$ is a discrete random variable with different values $y_1, y_2, \ldots$, then the $\sigma$-field $\mathcal{F}_Y$ is generated by the atoms

$$\{Y = y_1\}, \{Y = y_2\}, \ldots .$$

According to Definition 10.3, $E(X|Y)$ is constant on each of these atoms. It follows that the inverse image $\{E(X|Y) \in B\}$ of any Borel set $B$ in $\mathbb{R}$ is a countable union of some of these atoms, namely

$$\{E(X|Y) \in B\} = \bigcup_{i \in I} \{Y = i\},$$

where $I$ is the set of indices $i$ such that the value of $E(X|Y)$ on the atom $\{Y = i\}$ belongs to $B$. As a result, $\{E(X|Y) \in B\}$ belongs to $\mathcal{F}_Y$. This implies that $\mathcal{F}_{E(X|Y)} \subset \mathcal{F}_Y$.

**10.27** First, consider the case when $X \geqslant 0$. Let $Y$ be a discrete random variable with different values $y_1, y_2, \ldots$ . Then, $E(X|Y)$ is constant and equal to $E(X|\{Y = y_i\})$ on $\{Y = y_i\}$ for each $i$. It follows that

$$
\begin{aligned}
E(E(X|Y)) &= \sum_{i=1}^{\infty} E(X|\{Y = y_i\}) P(\{Y = y_i\}) \\
&= \sum_{i=1}^{\infty} \frac{E(\mathbb{I}_{\{Y=y_i\}} X)}{P(\{Y = y_i\})} P(\{Y = y_i\}) \\
&= \sum_{i=1}^{\infty} E(\mathbb{I}_{\{Y=y_i\}} X) = \lim_{n\to\infty} \sum_{i=1}^{n} E(\mathbb{I}_{\{Y=y_i\}} X) \\
&= \lim_{n\to\infty} E\left( \sum_{i=1}^{n} \mathbb{I}_{\{Y=y_i\}} X \right) = E(X)
\end{aligned}
$$

by Definition 9.7, since $\sum_{i=1}^{n} \mathbb{I}_{\{Y=y_i\}} X \nearrow X$ as $n \to \infty$.

Now, if $X$ is an arbitrary random variable, consider the positive and negative parts $X^+$ and $X^-$. Since $X^+, X^- \geqslant 0$ and $X = X^+ - X^-$, it follows that

$$
\begin{aligned}
E(E(X|Y)) &= E(E(X^+|Y)) + E(E(X^-|Y)) \\
&= E(X^+) + E(X^-) = E(X).
\end{aligned}
$$

**10.28** Let $Y$ be a discrete random variable with different values $y_1, y_2, \ldots$ . For any $i$, the random variable $E(X|Y)$ is constant on $\{Y = y_i\}$ and equal to $E(X|\{Y = y_i\})$. It follows that for each $i$,

$$
E(\mathbb{I}_{\{Y=y_i\}} E(X|Y)) = E(X|\{Y = y_i\}) P(\{Y = y_i\}) = E(\mathbb{I}_{\{Y=i\}} X).
$$

Now, take $B$ of the form

$$
B = \bigcup_{i \in I} \{Y = y_i\}, \tag{10.4}
$$

where $I \subset \{1, 2, \ldots\}$ is a countable set of indices, and put

$$
B_n = \bigcup_{i \in I_n} \{Y = y_i\},
$$

where $I_n = I \cap \{1, 2, \ldots, n\}$ is a finite set. This means that $B_n$ is a finite union of pairwise disjoint events of the form $\{Y = y_i\}$, so $\mathbb{I}_{B_n} = \sum_{i \in I_n} \mathbb{I}_{\{Y=y_i\}}$, which is a finite sum. As a result,

$$
\begin{aligned}
E(\mathbb{I}_{B_n} E(X|Y)) &= \sum_{i \in I_n} E(\mathbb{I}_{\{Y=y_i\}} E(X|Y)) \\
&= \sum_{i \in I_n} E(\mathbb{I}_{\{Y=y_i\}} X) = E(\mathbb{I}_{B_n} X).
\end{aligned}
$$

If $X \geqslant 0$, then $\mathbb{I}_{B_n}X \nearrow \mathbb{I}_B X$ and $\mathbb{I}_{B_n}E(X|Y) \nearrow \mathbb{I}_B E(X|Y)$ as $n \to \infty$, and monotone convergence (condition (c) in Definition 9.7) implies that $E(\mathbb{I}_{B_n}X) \nearrow E(\mathbb{I}_B X)$ and $E(\mathbb{I}_{B_n}E(X|Y)) \nearrow E(\mathbb{I}_B E(X|Y))$. It follows that

$$E(\mathbb{I}_B E(X|Y)) = E(\mathbb{I}_B X)$$

for any random variable $X \geqslant 0$. Finally, for an arbitrary random variable $X$, we can take the positive and negative parts $X^+, X^- \geqslant 0$ to get

$$\begin{aligned} E(\mathbb{I}_B E(X|Y)) &= E(\mathbb{I}_B E(X^+|Y)) - E(\mathbb{I}_B E(X^-|Y)) \\ &= E(\mathbb{I}_B X^+) - E(\mathbb{I}_B X^-) = E(\mathbb{I}_B X). \end{aligned}$$

This completes the proof because every $B \in \mathcal{F}_Y$ must be of the form (10.4).

**10.29** Since $Y$ is constant, $\mathcal{F}_Y = \{\emptyset, \Omega\}$. Thus, $E(X|Y)$ as a random variable with respect to this $\sigma$-field has to be constant too. We can determine the value of the latter constant from condition (b) in Definition 10.5 with $B = \Omega$, obtaining

$$\begin{aligned} E(X|Y) &= E(E(X|Y)) = E(\mathbb{I}_\Omega E(X|Y)) \\ &= E(\mathbb{I}_\Omega X) = E(X) = \int_0^1 2x\,\mathrm{d}x = 1; \end{aligned}$$

see Figure 10.1.

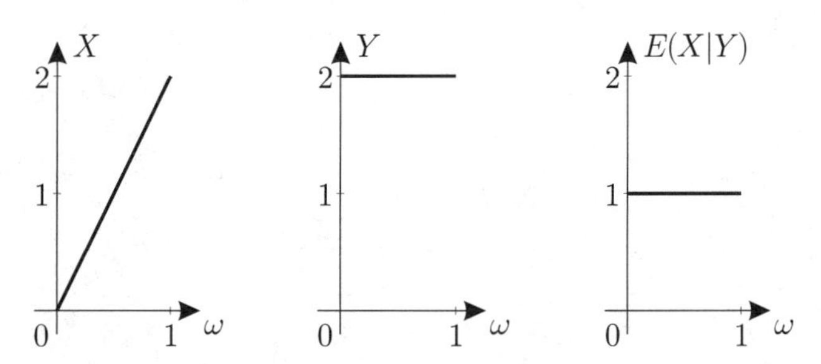


FIGURE 10.1. Random variables $X$, $Y$, and $E(X|Y)$ in Problem 10.29

**10.30** The $\sigma$-field $\mathcal{F}_Y$ has two atoms $B_1 = [0, \frac{1}{2})$ and $B_2 = [\frac{1}{2}, 1]$. Hence, $E(X|Y)$ is constant on $B_1$ and on $B_2$. The values $E(X|B_1)$ and $E(X|B_2)$ of these constants can be found from condition (b) in Definition 10.5. Since $B_1, B_2 \in \mathcal{F}_Y$,

$$E(X|B_1) = \frac{E(\mathbb{I}_{B_1}E(X|Y))}{P(B_1)} = \frac{E(\mathbb{I}_{B_1}X)}{P(B_1)} = \frac{1}{1/2}\int_0^{1/2} 2x\,\mathrm{d}x = \frac{1}{2},$$

$$E(X|B_2) = \frac{E(\mathbb{I}_{B_2}E(X|Y))}{P(B_2)} = \frac{E(\mathbb{I}_{B_2}X)}{P(B_2)} = \frac{1}{1/2}\int_{1/2}^{1} 2x\,\mathrm{d}x = \frac{3}{2}.$$

It follows that

$$E(X|Y)(\omega) = \begin{cases} 1/2 & \text{if } \omega \in [0, 1/2) \\ 3/2 & \text{if } \omega \in [1/2, 1]. \end{cases}$$

This is illustrated in Figure 10.2.

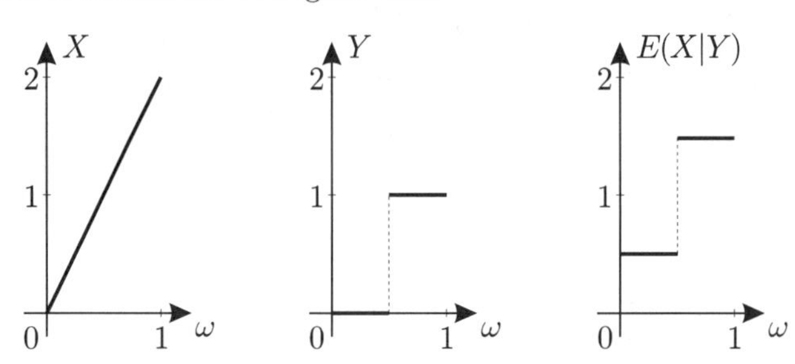


FIGURE 10.2. Random variables $X$, $Y$, and $E(X|Y)$ in Problem 10.30

**10.31** First, we show that $\mathcal{F}_Y$ contains all the Borel subsets in $[0, 1]$. To this end it is sufficient to show that all open intervals are in $\mathcal{F}_Y$. Take $(a, b) \subset [0, 1]$ and note that $(a, b) = \{Y \in (a^2, b^2)\} \in \mathcal{F}_Y$.

Because $[0, a) \in \mathcal{F}_Y$ for any $a \in [0, 1]$, we have

$$\int_0^a E(X|Y)(\omega)\, d\omega = \int_0^a X(\omega)\, d\omega.$$

If both $X$ and $E(X|Y)$ were continuous functions on $[0, 1]$, then differentiating both sides with respect to $a$, we would obtain

$$E(X|Y)(\omega) = X(\omega) = 2\omega$$

for each $\omega \in [0, 1]$. Even though we have no a priori guarantee that $E(X|Y)$ is continuous, it, nevertheless, turns out that the last equality is true. Indeed, $E(X|Y) = X$ clearly satisfies (a) of Definition 10.5, since $\mathcal{F}_Y$ is the $\sigma$-field consisting of all Borel subsets in $[0, 1]$, so $\mathcal{F}_X \subset \mathcal{F}_Y$. Moreover, $E(X|Y) = X$ satisfies (b) because the two expectations in (b) must obviously be equal if the random variables are the same. The graphs of $X$, $Y$, and $E(X|Y)$ are shown in Figure 10.3.

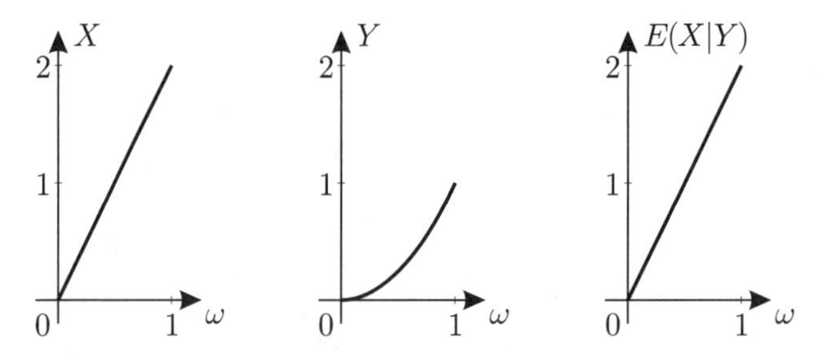


FIGURE 10.3. Random variables $X$, $Y$, and $E(X|Y)$ in Problem 10.31

**10.32** Since $\{Y = \frac{1}{2}\} = [\frac{1}{2}, 1]$, this event belongs to $\mathcal{F}_Y$. The inverse images of Borel sets contained in $[0, \frac{1}{2})$, which exhaust the family of all Borel subsets of $[0, \frac{1}{2})$, also belong to $\mathcal{F}_Y$. Now, each set in $\mathcal{F}_Y$ is of the form $B$ or $B \cup [\frac{1}{2}, 1]$, where $B$ is a Borel set contained in $[0, \frac{1}{2})$.

The random variable $E(X|Y)$ has to be constant on $[\frac{1}{2}, 1]$ or $\mathcal{F}_{E(X|Y)}$ would have to contain a proper Borel subset of $[\frac{1}{2}, 1]$, which is impossible if condition (a) of Definition 10.5 is to be satisfied. The value of $E(X|Y)$ on $[\frac{1}{2}, 1]$ is

$$E(X|[\tfrac{1}{2}, 1]) = \frac{1}{1/2} \int_{\frac{1}{2}}^{1} 2x \, \mathrm{d}x = \frac{3}{2}.$$

Next, let us find $E(X|Y)$ on $[0, \frac{1}{2})$. Because $[0, a)$ belongs to $\mathcal{F}_Y$ for any $a \in [0, \frac{1}{2})$, it follows by (b) of Definition 10.5 that

$$\int_0^a E(X|Y)(\omega) \, \mathrm{d}\omega = \int_0^a X(\omega) \, \mathrm{d}\omega.$$

If both $E(X|Y)$ and $X$ were continuous on $[0, \frac{1}{2})$, then we could differentiate both sides with respect to $a$ to obtain

$$E(X|Y)(\omega) = X(\omega) = 2\omega$$

for each $\omega \in [0, \frac{1}{2})$. Although we do not know in advance whether or not $E(X|Y)$ is indeed continuous on $[0, \frac{1}{2})$, this equality turns out to be true; namely we claim that

$$E(X|Y)(\omega) = \begin{cases} 2\omega & \text{if } \omega \in [0, 1/2) \\ 2/3 & \text{if } \omega \in [1/2, 1]; \end{cases}$$

see Figure 10.4. Indeed, condition (a) of Definition 10.5 holds because every set in $\mathcal{F}_{E(X|Y)}$ must be of the form $B$ or $B \cup [\frac{1}{2}, 1]$, where $B$ is a Borel set contained in $[0, \frac{1}{2})$, so $\mathcal{F}_{E(X|Y)} \subset \mathcal{F}_Y$. To verify condition (b), we take any set in $\mathcal{F}_Y$, which must be of the form $B$ or $B \cup [\frac{1}{2}, 1]$, where $B$ is a Borel set contained in $[0, \frac{1}{2})$. For any such $B$, the two expectations $E(\mathbb{I}_B E(X|Y))$ and $E(\mathbb{I}_B X)$ in (b) are equal because $E(X|Y) = X$ on $B$. For $B \cup [\frac{1}{2}, 1]$, we have

$$\begin{aligned} E(\mathbb{I}_{B\cup[1/2,1]} E(X|Y)) &= E(\mathbb{I}_B E(X|Y)) + E(\mathbb{I}_{[1/2,1]} E(X|Y)) \\ &= E(\mathbb{I}_B X) + P([\tfrac{1}{2}, 1]) E(X|[\tfrac{1}{2}, 1]) \\ &= E(\mathbb{I}_B X) + E(\mathbb{I}_{[1/2,1]} X) \\ &= E(\mathbb{I}_{B\cup[1/2,1]} X), \end{aligned}$$

completing the proof.

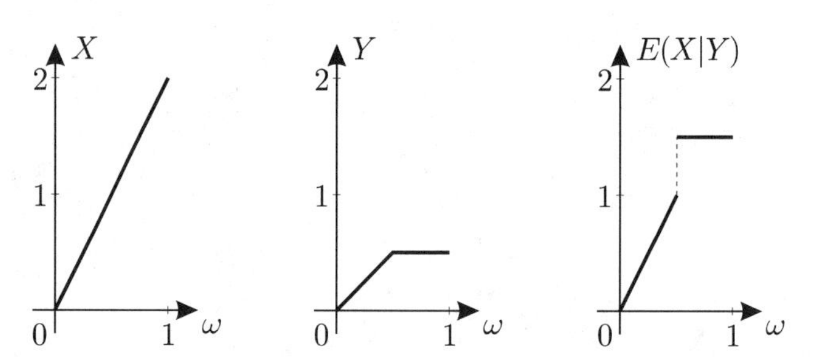


FIGURE 10.4. Random variables $X$, $Y$, and $E(X|Y)$ in Problem 10.32

**10.33** We claim that $\mathcal{F}_Y$ consists of all sets of the form $B\cup(1-B)$ such that $B \subset [0, \frac{1}{2}]$ is a Borel set. Indeed, if $A$ is any Borel set in $\mathbb{R}$, then $B = \{Y \in A\} \cap [0, \frac{1}{2}]$ is a Borel set in $[0, \frac{1}{2}]$ and $\{Y \in A\} = B \cup (1 - B)$. Conversely, if $B$ is any Borel set in $[0, \frac{1}{2}]$, then $A = 2 - 2B$ is a Borel set in $\mathbb{R}$, so $B \cup (1 - B) = \{Y \in A\}$ belongs to $\mathcal{F}_Y$. Our claim is verified.

We are ready to find $E(X|Y)$. Let us take any event $B\cup(1-B) \in \mathcal{F}_Y$, where $B \subset [0, \frac{1}{2}]$ is a Borel set. We put $\tilde{X}(\omega) = X(1-\omega)$. Then,

$$\begin{aligned} E(\mathbb{I}_{B\cup(1-B)}X) &= E(\mathbb{I}_B X) + E(\mathbb{I}_{1-B} X) \\ &= \frac{E(\mathbb{I}_B X) + E(\mathbb{I}_{1-B} X)}{2} + \frac{E(\mathbb{I}_{1-B}\tilde{X}) + E(\mathbb{I}_B \tilde{X})}{2} \\ &= \frac{E(\mathbb{I}_{B\cup(1-B)}X)}{2} + \frac{E(\mathbb{I}_{B\cup(1-B)}\tilde{X})}{2} \\ &= E\left(\mathbb{I}_{B\cup(1-B)}\frac{1}{2}(X+\tilde{X})\right). \end{aligned}$$

Because $\frac{1}{2}(X+\tilde{X})$ is a random variable with respect to $\mathcal{F}_Y$, it follows that

$$E(X|Y)(\omega) = \frac{X(\omega) + \tilde{X}(\omega)}{2} = \frac{X(\omega) + X(1-\omega)}{2} = |2\omega - 1|;$$

see Figure 10.5.

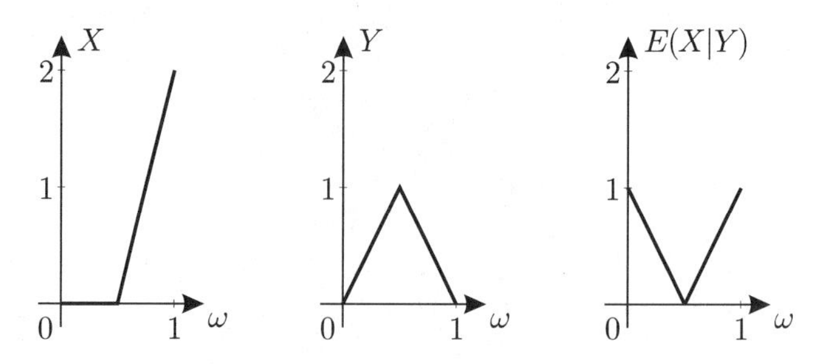


FIGURE 10.5. Random variables $X$, $Y$, and $E(X|Y)$ in Problem 10.33

**10.36** Apply condition (b) of Definition 10.5 with $B = \Omega$ to get

$$E(E(X|Y)) = E(\mathbb{I}_\Omega E(X|Y)) = E(\mathbb{I}_\Omega X) = E(X).$$

**10.37** By condition (a) in Definition 10.5, we have $\mathcal{F}_{E(U|Y)} \subset \mathcal{F}_Y$ for any random variable $U$. In particular, for $U = E(X|Z)$,

$$\mathcal{F}_{E(E(X|Z)|Y)} \subset \mathcal{F}_Y.$$

Next, let us take any event $B \in \mathcal{F}_Y$. Then, of course, $B \in \mathcal{F}_Z$, since $\mathcal{F}_Y \subset \mathcal{F}_Z$. We claim that

$$E(\mathbb{I}_B E(E(X|Z)|Y)) = E(\mathbb{I}_B E(X|Z)) = E(\mathbb{I}_B X).$$

The first equality holds because $B \in \mathcal{F}_Y$, so by condition (b) in Definition 10.5, we have $E(\mathbb{I}_B E(U|Y)) = E(\mathbb{I}_B U)$ for any random variable $U$, in particular, for $U = E(X|Z)$. The second equality holds by virtue of the same condition (b), since $B \in \mathcal{F}_Z$.

It follows that conditions (a) and (b) remain satisfied if every occurrence of $E(X|Y)$ is replaced by $E(E(X|Z)|Y)$ in Definition 10.5. By the uniquenness result in Theorem 10.1, it follows that

$$E(E(X|Z)|Y) = E(X|Y),$$

except perhaps on a set of probability zero.

**10.38** We can think of $E(X)$ as a constant random variable. Then, $\mathcal{F}_{E(X)} = \{\emptyset, \Omega\}$, so $\mathcal{F}_{E(X)} \subset \mathcal{F}_Y$ for any random variable $Y$.

Now, take any $B \in \mathcal{F}_Y$. If $X$ and $Y$ are independent random variables, then so are $X$ and $\mathbb{I}_B$. It follows that $E(\mathbb{I}_B)E(X) = E(\mathbb{I}_B X)$, so

$$E(\mathbb{I}_B E(X)) = E(\mathbb{I}_B)E(X) = E(\mathbb{I}_B X).$$

We have shown that conditions (a) and (b) remain satisfied if all occurrences of $E(X|Y)$ are replaced by $E(X)$ in Definition 10.5. By uniqueness (see Theorem 10.1), it follows that $E(X|Y) = E(X)$, except perhaps on a set of probability zero.

**10.39** We need to show that conditions (a) and (b) remain satisfied if every occurrence of $E(ZX|Y)$ is replaced by $ZE(X|Y)$ in Definition 10.5.

Condition (a) presents no difficulty: If $Z$ and $E(X|Y)$ are random variables with respect to the $\sigma$-field $\mathcal{F}_Y$, then so is the product $ZE(X|Y)$, which means that

$$\mathcal{F}_{ZE(X|Y)} \subset \mathcal{F}_Y.$$

To verify (b), we first take $Z = \mathbb{I}_C$ for any $C \in \mathcal{F}_Y$. Then, for any $B \in \mathcal{F}_Y$,

$$\begin{aligned} E(\mathbb{I}_B E(\mathbb{I}_C X|Y)) &= E(\mathbb{I}_B \mathbb{I}_C X) = E(\mathbb{I}_{B\cap C} X) \\ &= E(\mathbb{I}_{B\cap C} E(X|Y)) = E(\mathbb{I}_B \mathbb{I}_C E(X|Y)). \end{aligned}$$

The first equality holds by condition (b) in Definition 10.5, since $B \in \mathcal{F}_Y$. The third equality is true also by condition (b), since $B \cap C \in \mathcal{F}_Y$. The second and fourth equalities hold because $\mathbb{I}_B \mathbb{I}_C = \mathbb{I}_{B\cap C}$. It follows by uniqueness that $E(\mathbb{I}_C X|Y) = \mathbb{I}_C E(X|Y)$ (except perhaps on a set of probability zero).

Next, we verify (b) in the case of a random variable $Z$ with finitely many values $z_1, \dots, z_n$, which can be written as $Z = z_1 \mathbb{I}_{C_1} + \dots + z_n \mathbb{I}_{C_n}$ for some $C_1, \dots, C_n \in \mathcal{F}_Y$. Then,

$$\begin{aligned} E(ZX|Y) &= z_n E(\mathbb{I}_{C_1} X|Y) + \dots + z_n E(\mathbb{I}_{C_n} X|Y) \\ &= z_n \mathbb{I}_{C_1} E(X|Y) + \dots + z_n \mathbb{I}_{C_n} E(X|Y) = Z E(X|Y). \end{aligned}$$

Now, we shall verify (b) for any random variables $Z \geqslant 0$ and $X \geqslant 0$. We put

$$Z_n = \sum_{i=0}^{2^{n+1}} i 2^{-n} \mathbb{I}_{\{i \leqslant 2^n Z < i+1\}}.$$

Because $Z_n$ is a random variable with finitely many values such that $Z_n = \sum_{i=0}^{2^{n+1}} i 2^{-n} \mathbb{I}_{C_i}$, where $C_i = \{i \leqslant 2^n Z < i+1\} \in \mathcal{F}_Y$, we already know that

$$E(Z_n X|Y) = Z_n E(X|Y).$$

We observe that $Z_n \nearrow Z$ as $n \to \infty$. It follows by monotone convergence (condition (c) in Definition 9.7) that for any $B \in \mathcal{F}_Y$,

$$\begin{aligned} E(\mathbb{I}_B E(ZX|Y)) &= E(\mathbb{I}_B ZX) \\ &= \lim_{n\to\infty} E(\mathbb{I}_B Z_n X) = \lim_{n\to\infty} E(\mathbb{I}_B E(Z_n X|Y)) \\ &= \lim_{n\to\infty} E(\mathbb{I}_B Z_n E(X|Y)) = E(\mathbb{I}_B Z E(X|Y)), \end{aligned}$$

since $\mathbb{I}_B Z_n X \nearrow \mathbb{I}_B ZX$ and $\mathbb{I}_B Z_n E(X|Y) \nearrow \mathbb{I}_B Z E(X|Y)$ as $n \to \infty$. By uniqueness, we obtain

$$E(ZX|Y) = ZE(X|Y)$$

for any random variables $X, Z \geqslant 0$. Finally, for arbitrary random variables $Z$ and $X$, we consider their positive and negative parts to get

$$\begin{aligned}
&E(ZX|Y) \\
&= E(Z^+X^+|Y) - E(Z^+X^-|Y) - E(Z^-X^+|Y) + E(Z^-X^-|Y) \\
&= Z^+E(X^+|Y) - Z^+E(X^-|Y) - Z^-E(X^+|Y) + Z^-E(X^-|Y) \\
&= ZE(X|Y).
\end{aligned}$$

**10.40** We shall verify conditions (a) and (b) of Definition 10.5.

To show that (a) is satisfied, observe that $h(x|Y)$ is a random variable with respect to $\mathcal{F}_Y$ for each $x$, and the same is true for the right-hand side in (10.3).

To verify (b), put

$$g(y) = \int_{-\infty}^{+\infty} xh(x|Y)\,\mathrm{d}x$$

and take any $B$ in $\mathcal{F}_Y$. Then, there is a Borel set $A$ in $\mathbb{R}$ such that $B = \{Y \in A\}$ and

$$\begin{aligned}
E(\mathbb{I}_B E(X|Y)) = E(\mathbb{I}_A(Y)g(Y)) &= \int_{-\infty}^{+\infty} \mathbb{I}_A(y)g(y)f_Y(y)\,\mathrm{d}y \\
&= \int_{-\infty}^{+\infty}\int_{-\infty}^{+\infty} \mathbb{I}_A(y)xh(x|y)f_Y(y)\,\mathrm{d}x\,\mathrm{d}y \\
&= \int_{-\infty}^{+\infty}\int_{-\infty}^{+\infty} \mathbb{I}_A(y)xf_{X,Y}(x,y)\,\mathrm{d}x\,\mathrm{d}y \\
&= E(\mathbb{I}_A(Y)X) = E(\mathbb{I}_B X),
\end{aligned}$$

as required

**10.41** If $X$ and $Y$ are independent, then $f_{X,Y}(x,y) = f_X(x)f_Y(y)$ and

$$h(x|y) = \frac{f_{X,Y}(x,y)}{f_Y(y)} = f_X(x),$$

so

$$E(X|Y) = \int_{-\infty}^{+\infty} xh(x|Y)\,\mathrm{d}x = \int_{-\infty}^{+\infty} xf_X(x)\,\mathrm{d}x = E(X).$$

**10.42** First, we compute the density of $Y$,

$$f_Y(x) = \int_{-\infty}^{\infty} f_{X,Y}(x,y)\,\mathrm{d}x = K\int_0^1 (x+y)\,\mathrm{d}x = K(\tfrac{1}{2}+y).$$

Next, we find

$$h(x|y) = \frac{f_{X,Y}(x,y)}{f_Y(y)} = \frac{x+y}{y+1/2}$$

for $0 \leqslant x, y \leqslant 1$ and zero otherwise. Finally, we compute

$$E(X|Y) = \int_{-\infty}^{+\infty} x h(x|Y)\,\mathrm{d}x = \int_0^1 x \frac{x+Y}{1/2+Y}\,\mathrm{d}x = \frac{2+3Y}{3+3Y}.$$

There is no need to compute the constant $K$ explicitly.

**10.43** The density of $Y$ is

$$f_Y(x) = \int_{-\infty}^{\infty} f_{X,Y}(x,y)\,\mathrm{d}x = K \int_0^{\frac{\pi}{2}} \cos x \cos y\,\mathrm{d}x = K \cos y.$$

Next, we find

$$h(x|y) = \frac{f_{X,Y}(x,y)}{f_Y(y)} = \cos x$$

for $0 \leqslant x, y \leqslant \frac{\pi}{2}$ and zero otherwise. Finally,

$$E(X|Y) = \int_{-\infty}^{+\infty} x h(x|Y)\,\mathrm{d}x = \int_0^{\frac{\pi}{2}} x \cos x\,\mathrm{d}x = \frac{\pi}{2} - 1.$$

**10.44** First, we compute the density of $Y$,

$$\begin{aligned} f_Y(x) &= \int_{-\infty}^{\infty} f_{X,Y}(x,y)\mathrm{d}x \\ &= K \int_{-\sqrt{1-y^2}}^{\sqrt{1-y^2}} (2x+y)\,\mathrm{d}x = 2Ky\sqrt{1-y^2}. \end{aligned}$$

Next, we find

$$h(x|y) = \frac{f_{X,Y}(x,y)}{f_Y(y)} = \frac{2x+y}{2y\sqrt{1-y^2}}$$

if $x^2 + y^2 \leqslant 1$ and zero otherwise. Finally,

$$E(X,Y) = \int_{-\infty}^{+\infty} x h(x|Y)\,\mathrm{d}x = \int_{-\sqrt{1-Y^2}}^{\sqrt{1-Y^2}} \frac{2x+Y}{2Y\sqrt{1-Y^2}}\,\mathrm{d}x = 1.$$

**10.47** Clearly, $P(\{Y=i\}) = \frac{1}{3}$ for $i = 1, 2, 3$. We shall find $E(X|\{Y=i\})$ for each $i$.

If $i = 1$, then the player will avoid the trap if the first toss produces tails, and he will fall into the trap otherwise. Hence,

$$E(X|\{Y=1\}) = 10 \times \frac{1}{2} - 10 \times \frac{1}{2} = 0.$$

If $i = 2$, then the trap will be avoided if and only if the first toss produces heads and the second toss produces tails. Thus,

$$E(X|\{Y = 2\}) = 10 \times \frac{1}{4} - 10 \times \frac{3}{4} = -5.$$

If $i = 3$, then the player must get two tails or two heads and a tail in order to avoid the trap. As a result,

$$E(X|\{Y = 3\}) = 10 \times \frac{3}{8} - 10 \times \frac{5}{8} = -\frac{5}{2}.$$

It follows that

$$E(X|Y)(\omega) = \begin{cases} 0 & \text{if } \omega \in \{Y = 1\} \\ -5 & \text{if } \omega \in \{Y = 2\} \\ -5/2 & \text{if } \omega \in \{Y = 3\}. \end{cases}$$

Using the formula in Problem 10.36, we find

$$E(X) = E(E(X|Y)) = 0 \times \frac{1}{3} - 5 \times \frac{1}{3} - \frac{5}{2} \times \frac{1}{3} = -\frac{5}{2}.$$

**10.48** The values of $Y$ are $0, 1, 2, \ldots$ . The corresponding values of $E(X|Y)$ are

$$E(X|Y)(\omega) = E(X|\{Y = k\}) = kp \quad \text{if } \omega \in \{Y = k\}.$$

Using the formula in Problem 10.36, we can find the expectation of $X$,

$$E(X) = E(E(X|Y)) = \sum_{k=1}^{\infty} pkP(\{Y = k\}) = pE(Y) = p\lambda.$$

**10.49** To use the formula in Problem 10.36, we find that

$$E(X|N)(\omega) = E(X|\{N = k\}) = \sum_{i=1}^{k} E(Y_i) = km$$

for any $\omega \in \{N = k\}$, so

$$E(X) = E(E(X|N)) = \sum_{k=1}^{\infty} kmP(\{N = k\}) = mE(N) = mn.$$

**10.50** Let $Y$ be the value of $\lambda$ for the selected typesetter and let $X$ be the total number of misprints, $X = X_1 + X_2 + X_3$, where $X_i$ is the number of misprints on page $i$. The conditional expectation $E(X_i|Y)$ has four different values,

$$E(X_i|Y)(\omega) = E(X_i|\{Y = \lambda\}) = \lambda$$

for any $\omega \in \{Y = \lambda\}$, where $\lambda = 1, 2, 3, 4$. It follows that $E(X|Y) = E(X_1|Y) + E(X_2|Y) + E(X_3|Y)$ has four different values

$$E(X|Y)(\omega) = E(X|\{Y = \lambda\}) = 3\lambda,$$

for any $\omega \in \{Y = \lambda\}$, where $\lambda = 1, 2, 3, 4$. Finally,

$$E(X) = E(E(X|Y)) = \sum_{\lambda=1}^{4} 3\lambda P(\{Y = \lambda\}) = \sum_{\lambda=1}^{4} 3\lambda \frac{1}{4} = 7.5 \; .$$

# 11

# Characteristic Functions

## 11.1 Theory and Problems

In this chapter, we shall need random variables whose values are in the set of complex numbers $\mathbb{C}$. A random variable $Z$ with values in $\mathbb{C}$ can always be written as $X + iY$, where $X$ and $Y$ are $\mathbb{R}$-valued random variables, as in Definition 8.1, and where $i = \sqrt{-1}$ is the imaginary unit. The expectation of such a random variable $Z$ is a complex number defined by

$$E(Z) = E(X) + iE(Y).$$

For any random variable $X$ with values in $\mathbb{R}$ and for any $t \in \mathbb{R}$, by $e^{itX}$ we understand

$$e^{itX} = \cos(tX) + i\sin(tX).$$

In what follows, we shall study the expectation $E(e^{itX})$. The reason why this is extremely interesting is that $E(e^{itX})$ as a function of $t$ contains all the information to be found in the distribution of $X$. In many cases, the distribution of a random variable is awkward to deal with, whereas the expectation $E(e^{itX})$ is a well-behaved object.

▶ **Definition 11.1** Let $X$ be an arbitrary random variable. Then, $\phi_X : \mathbb{R} \to \mathbb{C}$ defined by

$$\phi_X(t) = E\left(e^{itX}\right) = E\left(\cos(tX)\right) + iE\left(\sin(tX)\right)$$

for all $t \in \mathbb{R}$ is called the *characteristic function* of $X$.

**11.1** Find the characteristic function of a random variable $X$ such that $P(\{X=1\}) = P(\{X=-1\}) = \frac{1}{2}$.

**11.2** Find the characteristic function of a random variable $X$ such that $P(\{X=1\}) = p$ and $P(\{X=0\}) = q$, where $p+q=1$ (Bernoulli distribution).

**11.3** Find the characteristic function of a random variable $X$ such that $P(\{X=k\}) = 2^{-k}$, $k=1,2,\ldots$ .

**11.4** Find the characteristic function of a random variable $X$ with Poisson distribution. Can you sketch the graph of $\phi_X(t)$?

**11.5** Find the characteristic function of a random variable $X$ with uniform distribution over the interval $[-1,1]$. Sketch the graph of this characteristic function.

**11.6** Find the characteristic function $\phi_X(t)$ of a random variable $X$ having distribution with density

$$f_X(x) = \begin{cases} 1-|x| & \text{if } |x| \leqslant 1 \\ 0 & \text{otherwise.} \end{cases}$$

**11.7** Find the characteristic function $\phi_X(t)$ of a random variable $X$ with distribution

$$P_X(A) = \frac{1}{2}\mathbb{I}_A(0) + \frac{1}{4}\int_{-1}^{1} \mathbb{I}_A(x)\,dx.$$

**11.8** What is the characteristic function $\phi_X(t)$ of a random variable $X$ with exponential distribution? Sketch the graph of $\phi_X(t)$.

**11.9** What is the characteristic function $\phi_X(t)$ of a normally distributed random variable $X$ with mean 0 and variance 1?

We shall now establish some general properties of the characteristic function.

**11.10** Show that for any random variable $X$

a) $\phi_X(0) = 1$,

b) $|\phi_X(t)| \leqslant 1$ for all $t \in \mathbb{R}$.

c) $\phi_X(-t) = \overline{\phi_X(t)}$ for all $t \in \mathbb{R}$.

(By $\bar{z}$ we denote the complex conjugate of $z \in \mathbb{C}$; that is, if $z = a+ib$, where $a,b \in \mathbb{R}$, then $\bar{z} = a - ib$.)

**11.11** Let $X$ and $Y$ be random variables such that $Y = aX + b$, where $a,b \in \mathbb{R}$. Show that

$$\phi_Y(t) = e^{itb}\phi_X(at).$$

**11.12** Use the formula in Problem 11.11 and the characteristic function of the normal distribution with mean 0 and variance 1 obtained in Problem 11.9 to find the characteristic function of the normal distribution with arbitrary mean $m$ and variance $\sigma^2 > 0$.

**11.13** Let $X$ be a random variable such that $P(\{X \in \mathbb{Z}\}) = 1$. Demonstrate that $\phi_X(t)$ is a periodic function with period $2\pi$; that is, $\phi_X(t + 2\pi) = \phi_X(t)$ for all $t$.

**11.14** Is it true that if $\phi_X(2\pi) = 1$, then $P(\{X \in \mathbb{Z}\}) = 1$?

**11.15** Let $X$ be a random variable and let $a \neq 0$. Show that the following conditions are equivalent:

(i) $\phi_X(a) = 1$;

(ii) $\phi_X$ is periodic with period $|a|$;

(iii) $P(\{\frac{a}{2\pi} X \in \mathbb{Z}\}) = 1$.

**11.16** If $X$ is a random variable with density, then $\phi_X(t) < 1$ for all $t \neq 0$.

**11.17** Let $X$ be a random variable taking values in a finite set. Verify that the moments $E(X^n)$ of $X$ are given by

$$E(X^n) = \frac{1}{i^n} \frac{d^n}{dt^n} \phi_X(0).$$

This useful result is, in fact, valid for any random variable $X$ as long as the $n$th moment $E(X^n)$ exits. The proof of the general case, which involves differentiation under the integral sign, is slightly technical and will be omitted.

▶ **Theorem 11.1** *Let $X$ be a random variable and $n$ a positive integer such that $E(X^n)$ exists. Then,*

$$E(X^n) = \frac{1}{i^n} \frac{d^n}{dt^n} \phi_X(0).$$

**11.18** Use the formula in Theorem 11.1 to obtain an expression for the variance of $X$ in terms of the characteristic function $\phi_X$ and its derivatives at zero.

**11.19** Use Theorem 11.1 to compute the expectation $E(X)$ and the second moment $E(X^2)$ of a random variable $X$ with Poisson distribution.

**11.20** Use Theorem 11.1 to compute the expectation $E(X)$ and the second moment $E(X^2)$ of a random variable $X$ with normal distribution.

We shall now learn how to reconstruct the distribution of a random variable from the characteristic function. Let us begin with the case of a random variable with integer values.

**11.21** Let $X$ be a random variable such that $P(\{X \in \mathbb{Z}\}) = 1$. Show that for all $n \in \mathbb{Z}$,

$$P(\{X = n\}) = \frac{1}{2\pi} \int_0^{2\pi} e^{-itn} \phi_X(t)\, dt.$$

**11.22** What is the distribution of a random variable $X$ with characteristic function

$$\phi_X(t) = \frac{1}{2} e^{-it} + \frac{1}{3} + \frac{1}{6} e^{2it} \ ?$$

**11.23** Find the distribution of a random variable $X$ with characteristic function

$$\phi_X(t) = \cos\left(\frac{t}{2}\right).$$

**11.24** Find the distribution of a random variable $X$ with characteristic function

$$\phi_X(t) = \frac{2}{3e^{it} - 1}.$$

The next theorem provides a condition in terms of the characteristic function for a random variable to be absolutely continuous. It makes it possible to reconstruct the density in terms of the characteristic function.

▶ **Theorem 11.2** *If $X$ is a random variable such that*

$$\int_{-\infty}^{\infty} |\phi_X(t)|\, dt < \infty,$$

*then $X$ has an absolutely continuous distribution with density*

$$f_X(x) = \frac{1}{2\pi} \int_{-\infty}^{\infty} e^{-itx} \phi_X(t)\, dt.$$

**11.25** Find the distribution of a random variable with characteristic function $e^{-|t|}$.

**11.26** What is the distribution of a random variable with characteristic function $e^{-t^2/2}$?

**11.27** Find the distribution of a random variable with characteristic function $\frac{1}{1+t^2}$.

Above we have seen instances of the following general result on reconstructing the distribution of a random variable from the characteristic function.

▶ **Theorem 11.3** *The distribution of a random variable $X$ is uniquely determined by the characteristic function $\phi_X$. The distribution function $F_X$ can be expresses in terms of $\phi_X$ as follows:*

$$F_X(y) - F_X(x) = \frac{1}{2\pi} \lim_{T\to\infty} \int_{-T}^{T} \frac{e^{-itx} - e^{-ity}}{it} \phi_X(t)\,dt$$

*at any points of continuity $x, y$ of $F_X$.*

**11.28** Find the distribution of a random variable with characteristic function $\frac{1}{3}e^{it\sqrt{2}} + \frac{2}{3}e^{it\sqrt{3}}$.

**11.29** What is the distribution of a random variable with characteristic function $\frac{1}{2}e^{-t^2/2} + \frac{1}{2}$ ?

We have seen in Problem 8.74 how to compute the distribution of the sum of independent random variables. However, this involves convolution, with which is not always easy to work. Characteristic functions provide an excellent tool to work with sums of independent random variables because they replace convolution by multiplication.

▶ **Theorem 11.4** *If $X$ and $Y$ are independent random variables, then*

$$\phi_{X+Y}(t) = \phi_X(t)\phi_Y(t).$$

**11.30** Prove Theorem 11.4.

**11.31** Use Theorem 11.4 to find the characteristic function of a random variable with the binomial distribution.

**11.32** Let $X$ and $Y$ be independent random variables with Poisson distribution. Verify that $X + Y$ is also a random variable with Poisson distribution. If $\lambda$ and $\mu$ are the parameters of $X$ and $Y$, respectively, what is the parameter of $X + Y$?

**11.33** Let $X$ and $Y$ be normally distributed independent random variables. Show that $X+Y$ is also normally distributed. If $X$ has mean $m$ and variance $\sigma^2$ and $Y$ has mean $\tilde{m}$ and variance $\tilde{\sigma}^2$, what is the mean and variance of $X + Y$?

## 11.2 Hints

**11.1** What are the possible values of $e^{itX}$ for the given $X$? What are the probabilities of these values? You need to answer these questions to find the expectation of $e^{itX}$.

If you do not feel comfortable working with complex-valued random variables, find the expectations of $\cos(tX)$ and $\sin(tX)$ and use them to express the characteristic function, as in Definition 11.1.

**11.2** See Problem 11.1.

**11.3** Observe that $e^{itX}$ is a discrete random variable because $X$ is. What are the values of $e^{itX}$ and their probabilities? To compute $E(e^{itX})$, multiply the values of $e^{itX}$ by the corresponding probabilities and add up. The sum will be that of a geometric series. Do you know a formula for computing this sum?

In the case in hand, you will find it easier to work with the complex-valued random variable $e^{itX}$, rather than with the two real-valued random variables $\cos(tX)$ and $\sin(tX)$.

**11.4** What are the values of $e^{itX}$ and their probabilities when $X$ has the Poisson distribution? The expectation $E\left(e^{itX}\right)$ will again be expressed as a series. To find the sum of this series, you will need the expansion $e^x = \sum_{k=1}^{\infty} \frac{1}{k!} x^k$ of the exponential function.

To sketch the graph of $\phi_X(t)$, observe that each of its points is given by three real numbers: $t$ and the real and imaginary parts of $\phi_X(t)$. The graph of $\phi_X(t)$ will be a curve in three dimensions.

**11.5** What is the density $f_X$ of a random variable $X$ uniformly distributed over an interval? Use Theorem 9.1 to express the expectation of $e^{itX}$ in terms of the density $f_X$. When evaluating the expectation, you may need to consider the case when $t = 0$ separately.

In this problem, $\phi_X(t)$ turns out to be real-valued, so the graph can be sketched on the plane.

**11.6** See Problem 11.5.

**11.7** How does one write the expectation of a random variable $X$ in terms of its distribution $P_X$? How does one write the expectation of $f(X)$ in terms of $P_X$ (here $f(x) = e^{itx}$)? Consider the cases where the distribution is a) discrete, b) with density $f_X$, and c) a combination of these.

**11.8** The density of the exponential distribution is given in Problem 9.38.

**11.9** You will need the formula

$$\frac{1}{\sqrt{2\pi}} \int_{-\infty}^{\infty} e^{-\frac{1}{2}(x-z)^2}\, dx = 1,$$

which holds for any $z \in \mathbb{C}$.

**11.10** All three properties follow immediately from the definition. In b), use the inequality $|E(Y)| \leqslant E(|Y|)$.

**11.11** All you need is the definition of the characteristic function and the fact that a constant factors out of the expectation.

**11.12** Can you find $a$ and $b$ such that $aX + b$ has the normal distribution with mean $m$ and variance $\sigma^2$ given that $X$ has the normal distribution with mean 0 and variance 1?

**11.13** For any discrete random variable $X$, it is possible to write $E\left(e^{itX}\right)$ as the sum (or series) over all the values taken by $X$ (see, for example, Problem 11.1 or 11.3). Show that each term in this series is periodic with period $2\pi$ if $X$ takes values in $\mathbb{Z}$.

**11.14** Consider the real part of $\phi_X(2\pi)$. It is also equal to 1. Now, use the fact that if $E(Y) = 0$ for a non-negative random variable $Y$, then $P(\{Y = 0\}) = 1$.

**11.15** Consider $Y = \frac{a}{2\pi}X$ and use Problems 11.11, 11.13, and 11.14.

**11.16** Use Problem 11.15. What is the probability that $\frac{t}{2\pi}X = k$ for any fixed $k \in \mathbb{Z}$ if $X$ is a random variable with density? What is the probability that $\frac{t}{2\pi}X \in \mathbb{Z}$?

**11.17** Write $E(e^{itX})$ in terms of the (finitely many) values of $X$ and their probabilities. Then, differentiate with respect to $t$ and put $t = 0$.

**11.18** How does one express the variance of $X$ in terms of the moments of $X$?

**11.19** All you need to do is to compute the derivatives of the characteristic function of a Poisson distributed random variable. This characteristic function was found in Solution 11.4.

**11.20** See Problem 11.19. The characteristic function of the normal distribution was obtained in Solution 11.12.

**11.21** Write $\phi_X(t)$ as a series with terms $e^{itk}P(\{X = k\})$ and integrate each term separately. Is the series of integrals convergent? What is the sum of it? Observe that the series is dominated by the convergent series with non-negative terms $P(\{X = k\})$.

**11.22** Observe that $\phi_X(t) = 1$ for some $t \neq 0$. This makes it possible to use Problems 11.15 and 11.21 to determine the distribution of $X$.

**11.23** Show that $P(\{2X \in \mathbb{Z}\}) = 1$, so you can use Problem 11.21.

**11.24** Because $P(\{X \in \mathbb{Z})\} = 1$ (why?), the formula in Problem 11.21 can be used, but the integral is somewhat tedious to evaluate. To simplify the integration, write $\phi_X(t)$ as $\sum_{n=-\infty}^{\infty} p_n e^{itn}$ and find the coefficients $p_n = P(\{X = n\})$ using the formula

$$\sum_{n=1}^{\infty} q^n = \frac{q}{1-q}, \qquad |q| < 1,$$

for the sum of a geometric series. Then, integrate the above series term by term.

**11.25** Use Theorem 11.2.

**11.26** Use Theorem 11.2. You will need the value of the integral

$$\int_{-\infty}^{\infty} e^{-(t+ix)^2/2}\,dt = \int_{-\infty}^{\infty} e^{-t^2/2}\,dt = \sqrt{2\pi}.$$

**11.27** Use Theorem 11.2. You will need the value of the Laplace integral

$$\int_0^{\infty} \frac{\cos(xt)}{1+t^2}\,dt = \frac{\pi}{2}e^{-x} \quad (x > 0).$$

**11.28** The characteristic function $\frac{1}{3}e^{it\sqrt{2}} + \frac{2}{3}e^{it\sqrt{3}}$ is not periodic (why?), so the method of finding the distribution based on Problem 11.15 is not applicable. Nor is the integral of the modulus of this function finite and Theorem 11.2 does not apply either. Use Theorem 11.3 instead. You will need to know that for any $a > 0$,

$$\lim_{T\to\infty} \int_{-T}^{T} \frac{\sin(\pm at)}{t}\,dt = \pm\pi.$$

**11.29** Use the formula in Theorem 11.3. To compute

$$\lim_{T\to\infty} \int_{-T}^{T} \frac{e^{-itx} - e^{-ity}}{it}\,dt,$$

see Solution 11.28. To compute

$$\lim_{T\to\infty} \int_{-T}^{T} \frac{e^{-itx} - e^{-ity}}{it} e^{-t^2/2}\,dt$$

observe that

$$\frac{e^{-itx} - e^{-ity}}{it} = \int_x^y e^{-itz}\,dt.$$

**11.30** If $X$ and $Y$ are independent random variables, then $e^{itX}$ and $e^{itY}$ are independent. What do you know about the expectation of the product of independent random variables?

**11.31** You can obtain a random variable with binomial distribution as the sum of independent random variables with Bernoulli distribution, the characteristic functions of which you already know.

**11.32** The characteristic function of a random variable with Poisson distribution is computed in Solution 11.4.

**11.33** The characteristic function of a random variable with the normal distribution is computed in Solution 11.12.

## 11.3 Solutions

**11.1** Since $X$ is a discrete random variable taking values 1 and $-1$ with probability $\frac{1}{2}$ each, $e^{itX}$ is also a discrete ($\mathbb{C}$-valued) random variable taking values $e^{it}$ and $e^{-it}$ with probability $\frac{1}{2}$ each. It follows that

$$\phi_X(t) = E\left(e^{itX}\right) = \frac{1}{2}e^{it} + \frac{1}{2}e^{-it} = \cos t.$$

**11.2** The random variable $e^{itX}$ takes two values $e^{it}$ and $1 = e^{it0}$ with probabilities $p$ and $q$, respectively. The characteristic function is therefore given by

$$\phi_X(t) = E\left(e^{itX}\right) = pe^{it} + q.$$

**11.3** The random variable $e^{itX}$ takes values $e^{itk}$ with probabilities $2^{-k}$, where $k = 1, 2, \ldots$ . It follows that

$$\begin{aligned}\phi_X(t) &= \sum_{k=1}^{\infty} 2^{-k}e^{itk} = \sum_{k=0}^{\infty}\left(\frac{e^{it}}{2}\right)^k - 1 = \frac{1}{1-\frac{e^{it}}{2}} - 1\\ &= \frac{e^{it}}{2-e^{it}}.\end{aligned}$$

Here, we have applied the formula

$$\sum_{k=0}^{\infty} a^k = \frac{1}{1-a},$$

valid whenever $|a| < 1$. Taking $a = e^{it}/2$, we have $|a| = 1/2 < 1$, as required, since $|e^{it}| = \sqrt{\cos^2 t + \sin^2 t} = 1$.

**11.4** If $X$ has the Poisson distribution with $P(\{X = k\}) = e^{-\lambda}\frac{\lambda^k}{k!}$, then $e^{itX}$ is a discrete random variable taking values $e^{itk}$ with probabilities $e^{-\lambda}\frac{\lambda^k}{k!}$ for $k = 0, 1, 2, \ldots$ . Therefore,

$$\begin{aligned}\phi_X(t) &= \sum_{k=0}^{\infty} e^{-\lambda}\frac{\lambda^k}{k!}e^{ikt} = e^{-\lambda}\sum_{k=0}^{\infty}\frac{\left(\lambda e^{it}\right)^k}{k!}\\ &= e^{-\lambda}e^{\lambda e^{it}} = e^{\lambda\left(e^{it}-1\right)}.\end{aligned}$$

Here, we have used the Maclaurin expansion

$$e^x = \sum_{k=0}^{\infty} \frac{x^k}{k!}$$

of the exponential function.

It is instructive to sketch the graph of the complex-valued function $\phi_X(t)$ as a curve in three dimensions, in the coordinate system formed by the real and imaginary axes and the $t$-axis; see Figure 11.1. To make it easier to visualize the line as a three-dimensional object, the picture shows a sausage whose central line is the graph of $\phi_X(t)$. We take $\lambda = 1$.

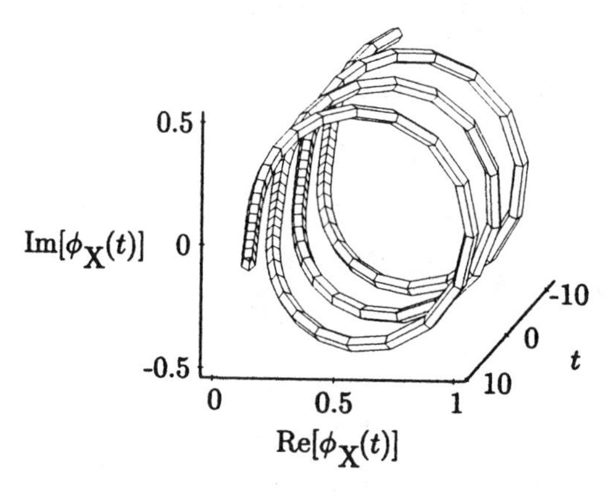


FIGURE 11.1. The characteristic function $\phi_X(t) = \exp(e^{it} - 1)$

**11.5** A random variable $X$ with uniform distribution over $[-1, 1]$ has density

$$f_X(x) = \begin{cases} \frac{1}{2} & \text{if } x \in [-1,1] \\ 0 & \text{otherwise.} \end{cases}$$

Thus,

$$\begin{aligned} \phi_X(t) = E\left(e^{itX}\right) &= \int_{-\infty}^{\infty} e^{itx} f_X(x)\, \mathrm{d}x \\ &= \frac{1}{2}\int_{-1}^{1} e^{itx}\, \mathrm{d}x = \frac{1}{2it} e^{itx}\bigg|_{x=-1}^{1} = \frac{1}{2it}\left(e^{it} - e^{-it}\right) \\ &= \frac{\sin t}{t}. \end{aligned}$$

This computation is valid provided that $t \neq 0$. For $t = 0$,

$$\phi_X(0) = E\left(e^0\right) = E(1) = 1.$$

In this case, $\phi_X(t)$ is a real-valued function, so the graph can be sketched on the plane; see Figure 11.2.

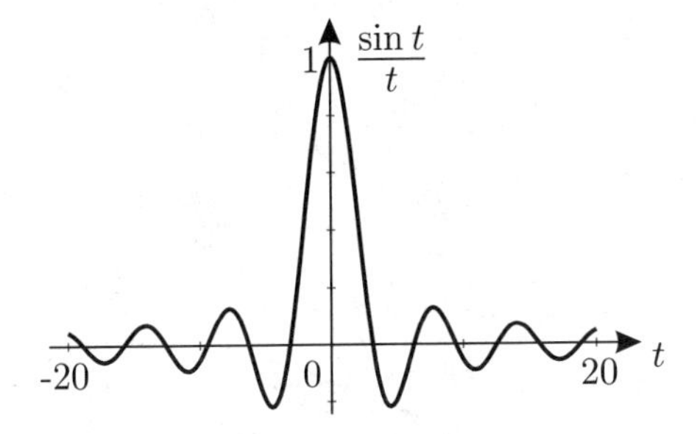


FIGURE 11.2. The characteristic function $\phi_X(t) = \frac{\sin t}{t}$

Observe that $\frac{\sin t}{t} \to 1$ as $t \to 0$. In this sense, the result $\phi_X(t) = \frac{\sin t}{t}$ applies for $t = 0$, too.

**11.6** When $t = 0$, we have $\phi_X(t) = 1$, as in Problem 11.5. If $t \neq 0$, then

$$\begin{aligned}\phi_X(t) = E\left(e^{itX}\right) &= \int_{-1}^{1} e^{itx} f_X(x)\,dx = \int_{-1}^{1} e^{itx}\left(1-|x|\right)\,dx \\ &= \int_{-1}^{0} e^{itx}(1+x)\,dx + \int_{0}^{1} e^{itx}(1-x)\,dx \\ &= \left(\frac{1-\cos t}{t^2} - i\frac{t-\sin t}{t^2}\right) + \left(\frac{1-\cos t}{t^2} + i\frac{t-\sin t}{t^2}\right) \\ &= \frac{2-2\cos t}{t^2}.\end{aligned}$$

The limit of this expression as $t \to 0$ is 1, and in this sense, it also covers the case of $t = 0$.

**11.7** Here, $P_X$ is a combination of a discrete distribution concentrated at 0 and the uniform distribution over $[-1, 1]$. It follows that for $t \neq 0$,

$$\begin{aligned}\phi_X(t) = E\left(e^{itX}\right) &= \frac{1}{2}e^{it0} + \frac{1}{4}\int_{-1}^{1} e^{itx}\,dx = \frac{1}{2} + \frac{1}{4it}\left(e^{it} - e^{-it}\right) \\ &= \frac{1}{2} + \frac{\sin t}{2t}.\end{aligned}$$

For $t = 0$, we have $\phi_X(t) = 1$.

**11.8** The density $f_X$ of the exponential distribution is given in Problem 9.38. It follows that

$$\begin{aligned}\phi_X(t) &= \int_{-\infty}^{\infty} e^{itx} f_X(x)\,dx = \lambda \int_{0}^{\infty} e^{itx}e^{-\lambda x}\,dx \\ &= -\frac{i\lambda}{t+i\lambda}\lim_{x\to\infty}\left(e^{itx}e^{-\lambda x} - 1\right) = \frac{i\lambda}{t+i\lambda}.\end{aligned}$$

The graph of $\phi_X(t)$ for $\lambda = 1$ is presented in Figure 11.3. There is an asymptote as $t \to \pm\infty$, which is also shown in the figure.

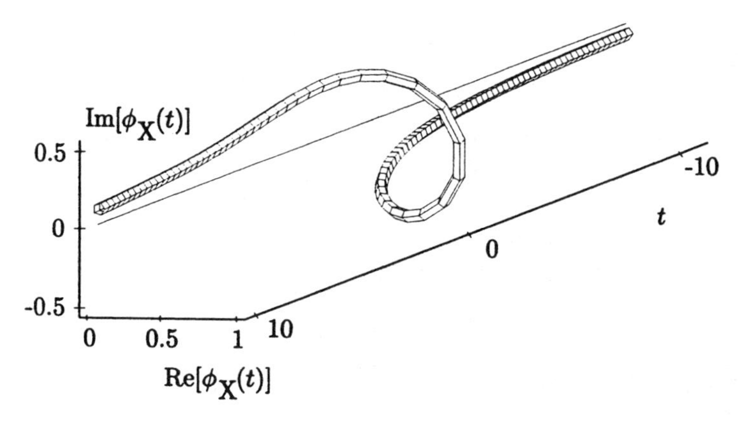


FIGURE 11.3. The characteristic function $\phi_X(t) = \frac{i}{t+i}$

**11.9** Since the density of the normal distribution with mean 0 and variance 1 is

$$f_X(x) = \frac{1}{\sqrt{2\pi}} e^{-\frac{1}{2}x^2},$$

we have

$$\begin{aligned}\phi_X(t) &= \int_{-\infty}^{\infty} e^{itx} f_X(x)\,\mathrm{d}x = \frac{1}{\sqrt{2\pi}} \int_{-\infty}^{\infty} e^{-\frac{1}{2}x^2+itx}\,\mathrm{d}x \\ &= \frac{1}{\sqrt{2\pi}} e^{-\frac{1}{2}t^2} \int_{-\infty}^{\infty} e^{-\frac{1}{2}(x-it)^2}\,\mathrm{d}x.\end{aligned}$$

However,

$$\frac{1}{\sqrt{2\pi}} \int_{-\infty}^{\infty} e^{-\frac{1}{2}(x-z)^2}\,\mathrm{d}x = 1$$

for any $z \in \mathbb{C}$, so

$$\phi_X(t) = e^{-\frac{1}{2}t^2}.$$

**11.10** a) $\phi_X(0) = E\left(e^0\right) = E(1) = 1.$
b) $|\phi_X(t)| = \left|E\left(e^{itX}\right)\right| \leqslant E\left(\left|e^{itX}\right|\right) = E(1) = 1.$
c) $\phi_X(-t) = E\big(e^{-itX}\big) = E\big(\overline{e^{itX}}\big) = \overline{E\big(e^{itX}\big)} = \overline{\phi_X(t)}.$

**11.11** From the definition of the characteristic function

$$\begin{aligned}\phi_Y(t) &= E\left(e^{itY}\right) = E\left(e^{it(aX+b)}\right) = E\left(e^{itb}e^{iatX}\right) \\ &= e^{itb}E\left(e^{iatX}\right) = e^{itb}\phi_X(at)\end{aligned}$$

if $Y = aX + b$.

**11.12** Suppose that $X$ has the normal distribution with mean $E(X) = 0$ and variance $V(X) = 1$. Then,

$$Y = \sigma X + m$$

has the normal distribution with mean $m$ and variance $\sigma^2$. Indeed,

$$\begin{aligned} E(Y) &= \sigma E(X) + m = m, \\ V(Y) &= E\left((Y - E(Y))^2\right) = \sigma^2 E\left(X^2\right) = \sigma^2 V(X) = \sigma^2. \end{aligned}$$

By Problems 11.11 and 11.9,

$$\phi_Y(t) = e^{itm}\phi_X(\sigma t) = e^{-\frac{1}{2}\sigma^2 t^2 + imt}.$$

**11.13** We put $p_k = P(\{X = k\})$ for each $k \in \mathbb{Z}$. Then,

$$\sum_{k=-\infty}^{\infty} p_k = \sum_{k=-\infty}^{\infty} P(\{X = k\}) = P(\{X \in \mathbb{Z}\}) = 1$$

and

$$\begin{aligned} \phi_X(t + 2\pi) &= \sum_{k=-\infty}^{\infty} e^{i(t+2\pi)k} p_k = \sum_{k=-\infty}^{\infty} \left(e^{2\pi i}\right)^k e^{itk} p_k \\ &= \sum_{k=-\infty}^{\infty} e^{itk} p_k = \phi_X(t), \end{aligned}$$

since $e^{2\pi i} = 1$.

**11.14** If

$$\phi_X(2\pi) = E\left(e^{2\pi i X}\right) = E\left(\cos(2\pi X)\right) + iE\left(\sin(2\pi X)\right) = 1,$$

then $E(\cos(2\pi X)) = 1$ or, equivalently, $E(1 - \cos(2\pi X)) = 0$. But $1 - \cos(2\pi X) \geqslant 0$, so $P(\{1 - \cos(2\pi X) = 0\}) = 1$. This means that $P(\{X \in \mathbb{Z}\}) = 1$. The statement in Problem 11.14 is true.

**11.15** Put $Y = \frac{a}{2\pi} X$. Then, $\phi_Y(t) = \phi_X\left(\frac{a}{2\pi} t\right)$ by Problem 11.11. We shall prove the following implications: (i) $\Rightarrow$ (iii) $\Rightarrow$ (ii) $\Rightarrow$ (i).

(i) $\Rightarrow$ (iii). Since $\phi_Y(2\pi) = \phi_X(a) = 1$, it follows by Problem 11.14 that $P(\{Y \in \mathbb{Z}\}) = 1$.

(iii) $\Rightarrow$ (ii). If $P(\{Y \in \mathbb{Z}\}) = 1$, then $\phi_Y(t)$ is periodic with period $2\pi$ by 11.13. But $\phi_X(t) = \phi_Y\left(\frac{2\pi}{a} t\right)$ by Problem 11.11, so $\phi_X(t)$ is periodic with period $|a|$.

(ii) $\Rightarrow$ (i). Since $\phi_X(0) = 1$ (see 11.10) and $\phi_X(t)$ is periodic with period $|a|$, it follows that $\phi_X(a) = \phi_X(0) = 1$.

**11.16** Let $t \neq 0$. By Problem 11.15, it suffices to show that $P(\{\frac{t}{2\pi}X \in \mathbb{Z}\}) < 1$ if $X$ is a random variable with density $f_X$. In fact, $P(\{\frac{t}{2\pi}X \in \mathbb{Z}\}) = 0$ in this case. Indeed, for any $k \in \mathbb{Z}$,

$$P(\{\tfrac{t}{2\pi}X = k\}) = \int_{2\pi k/t}^{2\pi k/t} f_X(x)\,\mathrm{d}x = 0$$

by the properties of integrals. It follows that

$$P(\{\tfrac{t}{2\pi}X \in \mathbb{Z}\}) = \sum_{k\in\mathbb{Z}} P(\{\tfrac{t}{2\pi}X = k\}) = 0.$$

**11.17** Suppose that $X$ can take $m$ values $x_1, \ldots, x_m$ with probabilities $p_1, \ldots, p_m$, respectively. Then, the random variable $e^{itX}$ will also have $m$ values $e^{itx_1}, \ldots, e^{itx_m}$ with the same probabilities, so

$$\phi_X(t) = E\left(e^{itX}\right) = \sum_{k=1}^{m} p_k e^{itx_k}.$$

We differentiate $n$ times with respect to $t$ to obtain

$$\frac{d^n}{dt^n}\phi_X(t) = \sum_{k=1}^{m} p_k (ix_k)^n e^{itx_k}.$$

Putting $t = 0$, we find that

$$\frac{d^n}{dt^n}\phi_X(0) = i^n \sum_{k=1}^{m} p_k x_k^n = i^n E(X^n),$$

as required.

**11.18** Since $V(X) = E(X^2) - E(X)^2$, we find by Theorem 11.1 that

$$V(X) = -\phi_X''(0) + \phi_X'(0)^2.$$

Since $\phi_X(0) = 1$, after a few transformations we obtain

$$V(X) = -(\ln \phi_X)''(0).$$

**11.19** In Solution 11.4, it was found that $\phi_X(t) = e^{\lambda(e^{it}-1)}$, so

$$\frac{d}{dt}\phi_X(t) = i\lambda e^{it} e^{\lambda(e^{it}-1)},$$

$$\frac{d^2}{dt^2}\phi_X(t) = i^2\lambda\left(1 + \lambda e^{2it}\right)e^{\lambda(e^{it}-1)}.$$

By Theorem 11.1,

$$E(X) = \frac{1}{i}\frac{d}{dt}\phi_X(0) = \lambda,$$

$$E(X^2) = \frac{1}{i^2}\frac{d^2}{dt^2}\phi_X(0) = \lambda(\lambda+1).$$

**11.20** If $X$ is a normally distributed random variable, then

$$\phi_X(t) = e^{-\frac{1}{2}\sigma^2 t^2 + imt}$$

by Solution 11.12. The derivatives with respect to $t$ are

$$\begin{aligned}
\frac{d}{dt}\phi_X(t) &= \left(-\sigma^2 t + im\right) e^{-\frac{1}{2}\sigma^2 t^2 + imt}, \\
\frac{d^2}{dt^2}\phi_X(t) &= \left(-\sigma^2 + \left(-\sigma^2 t + im\right)^2\right) e^{-\frac{1}{2}\sigma^2 t^2 + imt}.
\end{aligned}$$

Substituting $t = 0$, we obtain

$$\begin{aligned}
E(X) &= \frac{1}{i}\frac{d}{dt}\phi_X(0) = m, \\
E(X^2) &= \frac{1}{i^2}\frac{d^2}{dt^2}\phi_X(0) = \sigma^2 + m^2.
\end{aligned}$$

**11.21** We put $p_k = P(\{X = k\})$ for brevity. Then,

$$\phi_X(t) = \sum_{k\in\mathbb{Z}} e^{itk} p_k.$$

For any $n \in \mathbb{Z}$, let us multiply the $k$th term of this series by $e^{-itn}$ and integrate the result from 0 to $2\pi$:

$$\int_0^{2\pi} e^{-itn} e^{itk} p_k \, \mathrm{d}t = \int_0^{2\pi} e^{it(k-n)} p_k \, \mathrm{d}t = \begin{cases} 0 & \text{if } k \neq n \\ 2\pi p_n & \text{if } k = n. \end{cases}$$

If the sum and integral signs below can be interchanged, then

$$\begin{aligned}
\int_0^{2\pi} e^{-itn} \phi_X(t) \, \mathrm{d}t &= \int_0^{2\pi} \sum_{k\in\mathbb{Z}} e^{it(k-n)} p_k \, \mathrm{d}t \\
&= \sum_{k\in\mathbb{Z}} \int_0^{2\pi} e^{it(k-n)} p_k \, \mathrm{d}t = 2\pi p_n,
\end{aligned}$$

proving the formula in Problem 11.21.

Here is an argument to show that the sum and integral signs can be interchanged. Put

$$\begin{aligned}
a &= \int_0^{2\pi} \sum_{k\in\mathbb{Z}} e^{it(k-n)} p_k \, \mathrm{d}t, \\
a_N &= \sum_{k=-N}^{N} \int_0^{2\pi} e^{it(k-n)} p_k \, \mathrm{d}t = \int_0^{2\pi} \sum_{k=-N}^{N} e^{it(k-n)} p_k \, \mathrm{d}t
\end{aligned}$$

(integrals and finite sums can obviously be interchanged). We need to show that $a_N \to a$ as $N \to \infty$. This is so because

$$\begin{aligned} |a_N - a| &= \left| \int_0^{2\pi} \sum_{|k|>N} e^{it(k-n)} p_k \, dt \right| \\ &\leqslant \int_0^{2\pi} \sum_{|k|>N} \left| e^{it(k-n)} p_k \right| dt \\ &= 2\pi \sum_{|k|>N} p_k \to 0 \quad \text{as } N \to \infty, \end{aligned}$$

since $\sum_{k\in\mathbb{Z}} p_k = 1$ is a convergent series.

**11.22** Observe that $\phi_X(t)$ is periodic with period $2\pi$. This means that $P(\{X \in \mathbb{Z}\}) = 1$ by Problem 11.15. To determine the distribution of $X$, we need to find $p_n = P(\{X = n\})$ for all $n \in \mathbb{Z}$. The $p_n$ are given by the formula in Problem 11.21. For $n = -1$,

$$p_{-1} = \frac{1}{2\pi} \int_0^{2\pi} e^{it} \phi_X(t) \, dt = \frac{1}{2}.$$

Similarly, $p_0 = \frac{1}{3}$, $p_2 = \frac{1}{6}$, and $p_n = 0$ for any $n \in \mathbb{Z}$ other than $-1$, 0, or 2.

**11.23** Since $\phi_X(t) = \cos(t/2)$ is periodic with period $4\pi$, it follows that for $Y = 2X$, we have $P(\{Y \in \mathbb{Z}\}) = 1$ by Problem 11.15 and $\phi_Y(t) = \cos t$ by Problem 11.11. Using the formula in Problem 11.21 for $n = 1$, we find that

$$\begin{aligned} P\left(\left\{X = \tfrac{1}{2}\right\}\right) = P(\{Y = 1\}) &= \frac{1}{2\pi} \int_0^{2\pi} e^{it} \phi_Y(t) \, dt \\ &= \frac{1}{2\pi} \int_0^{2\pi} e^{it} \cos t \, dt = \frac{1}{2}. \end{aligned}$$

Similarly, for $n = -1$,

$$P\left(\left\{X = -\tfrac{1}{2}\right\}\right) = P(\{Y = -1\}) = \frac{1}{2}$$

and for any $n \notin \{-1, 1\}$,

$$P\left(\left\{X = \tfrac{n}{2}\right\}\right) = P(\{Y = n\}) = 0.$$

**11.24** The characteristic function $\phi_X(t)$ is periodic with period $2\pi$, so $P(\{X \in \mathbb{Z})\} = 1$ by Problem 11.15. Using the formula for the sum of a geometric series (see the hint), we can write the characteristic function as

$$\phi_X(t) = \frac{2}{3e^{it} - 1} = 2\frac{e^{-it}/3}{1 - e^{-it}/3} = 2\sum_{n=1}^{\infty}\left(\frac{e^{-it}}{3}\right)^n = \sum_{n=1}^{\infty}\frac{2}{3^n}e^{-itn}.$$

Now, the formula in Problem 11.21 gives

$$P(\{X = n\}) = \begin{cases} 0 & \text{if } n = 0, 1, 2, \ldots \\ 2 \times 3^n & \text{if } n = -1, -2, \ldots . \end{cases}$$

**11.25** Let $X$ be a random variable with characteristic function $\phi_X(t) = e^{-|t|}$. First, we verify that

$$\int_{-\infty}^{\infty} |\phi_X(t)|\, \mathrm{d}t = \int_{-\infty}^{\infty} e^{-|t|}\, \mathrm{d}t = 2\int_0^{\infty} e^{-t}\, \mathrm{d}t = 2 < \infty.$$

We can therefore use Theorem 11.2 to find the density of $X$,

$$\begin{aligned} f_X(x) &= \frac{1}{2\pi}\int_{-\infty}^{\infty} e^{-itx}\phi_X(t)\, \mathrm{d}t = \frac{1}{2\pi}\int_{-\infty}^{\infty} e^{-itx}e^{-|t|}\, \mathrm{d}t \\ &= \frac{1}{\pi}\int_0^{\infty} \cos(tx)e^{-t}\, \mathrm{d}t = \frac{1}{\pi\left(1 + x^2\right)}. \end{aligned}$$

This means that $X$ has the Cauchy distribution.

**11.26** Let $X$ be a random variable with characteristic function $\phi_X(t) = e^{-t^2/2}$. It is clear that

$$\int_{-\infty}^{\infty} |\phi_X(t)|\, \mathrm{d}t = \int_{-\infty}^{\infty} e^{-t^2/2}\, \mathrm{d}t = \sqrt{2\pi} < \infty.$$

By Theorem 11.2, $X$ is an absolutely continuous random variable with density

$$\begin{aligned} f_X(x) &= \frac{1}{2\pi}\int_{-\infty}^{\infty} e^{-itx}\phi_X(t)\, \mathrm{d}t = \frac{1}{2\pi}\int_{-\infty}^{\infty} e^{-itx}e^{-t^2/2}\, \mathrm{d}t \\ &= \frac{1}{2\pi}e^{-x^2/2}\int_{-\infty}^{\infty} e^{-(t+ix)^2/2}\, \mathrm{d}t = \frac{1}{\sqrt{2\pi}}e^{-x^2/2}. \end{aligned}$$

It follows that $X$ has the normal distribution with mean 0 and variance 1.

**11.27** Let $X$ be a random variable with characteristic function $\phi_X(t) = \frac{1}{1+t^2}$. We shall use Theorem 11.2 to find the distribution of $X$. First, we verify that

$$\int_{-\infty}^{\infty} |\phi_X(t)|\, \mathrm{d}t = \int_{-\infty}^{\infty} \frac{1}{1+t^2}\, \mathrm{d}t = \pi < \infty.$$

It follows that $X$ has an absolutely continuous distribution with density

$$\begin{aligned} f_X(x) &= \frac{1}{2\pi}\int_{-\infty}^{\infty} e^{-itx}\phi_X(t)\,\mathrm{d}t = \frac{1}{2\pi}\int_{-\infty}^{\infty}\frac{e^{-itx}}{1+t^2}\,\mathrm{d}t \\ &= \frac{1}{\pi}\int_0^{\infty}\frac{\cos(tx)}{1+t^2}\,\mathrm{d}t = \frac{1}{2}e^{-|x|}, \end{aligned}$$

where we have used the value of the Laplace integral given in the hint. The distribution with density $f_X(x) = \frac{1}{2}e^{-|x|}$ is known as the *Laplace distribution*.

**11.28** Using the limit in the hint, we find that for any $a \in \mathbb{R}$,

$$\begin{aligned} &\lim_{T\to\infty}\int_{-T}^{T}\frac{e^{-itx}-e^{-ity}}{it}e^{ita}\,\mathrm{d}t \\ &= \lim_{T\to\infty}\int_{-T}^{T}\frac{e^{-it(x-a)}-e^{-it(y-a)}}{it}\,\mathrm{d}t \\ &= \lim_{T\to\infty}\int_{-T}^{T}\frac{\sin[(y-a)t]-\sin[(x-a)t]}{t}\,\mathrm{d}t \\ &= \begin{cases} 0 & \text{if } x, y > a \text{ or } x, y < a \\ 2\pi & \text{if } x < a < y. \end{cases} \end{aligned}$$

(Other cases are possible, for example, $y < a < x$, but they will not be needed.) By Theorem 11.3,

$$\begin{aligned} &F_X(y) - F_X(x) \\ &= \frac{1}{2\pi}\lim_{T\to\infty}\int_{-T}^{T}\frac{e^{-itx}-e^{-ity}}{it}\left(\frac{1}{3}e^{it\sqrt{2}} + \frac{2}{3}e^{it\sqrt{3}}\right)\mathrm{d}t, \end{aligned}$$

so

$$F_X(y) - F_X(x) = \begin{cases} 0 & \text{if } x, y < \sqrt{2} \text{ or } \sqrt{3} < x, y \\ & \quad \text{or } \sqrt{2} < x, y < \sqrt{3} \\ 1/3 & \text{if } x < \sqrt{2} < y \\ 2/3 & \text{if } x < \sqrt{3} < y. \end{cases}$$

This means that $X$ has a discrete distribution with two values $\sqrt{2}$ and $\sqrt{3}$ and corresponding probabilities $\frac{1}{3}$ and $\frac{2}{3}$.

**11.29** Let $X$ be a random variable with characteristic function $\phi_X(t) = \frac{1}{2}e^{-t^2/2} + \frac{1}{2}$. To use the formula in Theorem 11.3, we need to compute the following limits:

$$\lim_{T\to\infty}\int_{-T}^{T}\frac{e^{-itx}-e^{-ity}}{it}\,\mathrm{d}t = \begin{cases} 0 & \text{if } x, y > 0 \text{ or } x, y < 0 \\ 2\pi & \text{if } x < 0 < y, \end{cases}$$

as in Solution 11.28, and

$$\lim_{T\to\infty}\int_{-T}^{T}\frac{e^{-itx}-e^{-ity}}{it}e^{-t^2/2}\,\mathrm{d}t = \lim_{T\to\infty}\int_{-T}^{T}\int_{x}^{y}e^{-itz}e^{-t^2/2}\,\mathrm{d}z\,\mathrm{d}t$$
$$= \lim_{T\to\infty}\int_{x}^{y}\int_{-T}^{T}e^{-itz}e^{-t^2/2}\,\mathrm{d}t\,\mathrm{d}z$$
$$= \sqrt{2\pi}\int_{x}^{y}e^{-z^2/2}\,\mathrm{d}z.$$

As a result,

$$F_X(y)-F_X(x)$$
$$=\frac{1}{2\pi}\lim_{T\to\infty}\int_{-T}^{T}\frac{e^{-itx}-e^{-ity}}{it}\left(\frac{1}{2}e^{-t^2/2}+\frac{1}{2}\right)\mathrm{d}t$$
$$=\frac{1}{2}\int_{x}^{y}\frac{e^{-z^2/2}}{\sqrt{2\pi}}\,\mathrm{d}z+\begin{cases}0 & \text{if } x,y>0 \text{ or } x,y<0\\ 1/2 & \text{if } x<0<y.\end{cases}$$

The distribution of $X$ is a mixture of the normal distribution with mean 0 and variance 1 and a discrete distribution, namely the Dirac measure concentrated at 0.

**11.30** Because $X$ and $Y$ are independent random variables, $e^{itX}$ and $e^{itY}$ are also independent; see Problem 8.69. Since the expectation of the product of independent random variables is the product of their expectations,

$$\phi_{X+Y}(t) = E\left(e^{it(X+Y)}\right) = E\left(e^{itX}e^{itY}\right) = E\left(e^{itX}\right)E\left(e^{itY}\right)$$
$$= \phi_X(t)\phi_Y(t).$$

**11.31** Let us consider $n$ independent random variables $X_1,\dots,X_n$, all having the same Bernoulli distribution with parameter $p$; that is, $P(\{X_k=1\})=p$ and $P(\{X_k=0\})=q=1-p$ for all $k=1,\dots,n$. Then, $X=X_1+\cdots+X_n$ has the binomial distribution $B(n,p)$. Since $\phi_{X_k}(t)=pe^{it}+q$ for each $k$, it follows by Theorem 11.4 that

$$\phi_X(t)=\phi_{X_1}(t)\cdots\phi_{X_n}(t)=\left(pe^{it}+q\right)^n.$$

**11.32** Let $X$ and $Y$ have the Poisson distribution with parameter $\lambda$ and $\mu$, respectively. If $X$ and $Y$ are independent, then

$$\phi_{X+Y}(t)=\phi_X(t)\phi_Y(t)=e^{\lambda(e^{it}-1)}e^{\mu(e^{it}-1)}=e^{(\lambda+\mu)(e^{it}-1)}.$$

It follows by the uniqueness asserted in Theorem 11.3 that $X+Y$ has the Poisson distribution with parameter $\lambda+\mu$.

**11.33** Suppose that $X$ and $Y$ have the normal distribution with mean $m$ and $\tilde{m}$ and variance $\sigma^2$ and $\tilde{\sigma}^2$, respectively. If $X$ and $Y$ are independent, then

$$\begin{aligned}\phi_{X+Y}(t) &= \phi_X(t)\phi_Y(t)\\ &= e^{-\frac{1}{2}\sigma^2 t^2 + imt} e^{-\frac{1}{2}\tilde{\sigma}^2 t^2 + i\tilde{m}t} = e^{-\frac{1}{2}(\sigma^2+\tilde{\sigma}^2)t^2 + i(m+\tilde{m})t}.\end{aligned}$$

It follows that $X+Y$ has the normal distribution with mean $m+\tilde{m}$ and variance $\sigma^2 + \tilde{\sigma}^2$.

# 12
# Limit Theorems

## 12.1 Theory and Problems

This chapter is devoted to problems concerned with the asymptotic behavior of sums of many independent identically distributed random variables. The results constitute an important part of probability theory because of their theoretical significance and practical consequences. Some proofs tend to be quite difficult, however, and are deferred to a more advanced course in probability. Instead, the consequences and applications will be highlighted here.

Nevertheless, the main points can be seen even in the relatively straightforward *weak law of large numbers*, which will be introduced in detail through a series of problems below. Then, we shall state two major results, the *strong law of large numbers* and *central limit theorem*, and study their consequences.

**12.1** An archer is aiming at a circular target of radius 20 inches. Her arrows hit on average 5 inches away from the center, each shot being independent. Show that the next arrow will miss the target with probability at most $\frac{1}{4}$.

The archer is likely to miss the target no more than once per four arrows shot. Possibly, she will do much better than that. But because all we know here is the average distance between the point hit and the center of the target, we are unable to obtain a sharper estimate.

$*$ **12.2** Show that $\frac{1}{4}$ is the best possible estimate of the probability of missing the target in Problem 12.1; that is, in certain circumstances, the probability can be equal to or be arbitrarily close to $\frac{1}{4}$.

Problem 12.1 leads to a general property, whose consequences reach far beyond the noble art of archery.

▶ **12.3** Show that if a non-negative random variable has finite expectation $E(X)$, then

$$P(\{X \geqslant C\}) \leqslant \frac{E(X)}{C}$$

for any $C > 0$.

The inequality in Theorem 12.1 is a straightforward consequence of that in Problem 12.3.

**12.4** Prove the following theorem, known as the *Chebyshev inequality.*

▶ **Theorem 12.1** *Let $X$ be a random variable with finite expectation $E(X)$ and finite variance $V(X)$. Then,*

$$P(\{|X - E(X)| \geqslant C\}) \leqslant \frac{V(X)}{C^2}$$

*for any $C > 0$.*

The Chebyshev inequality provides a very rough estimate, whose strength comes from its universality. It allows us to estimate the likelihood of a deviation of a random variable $X$ from its mean value $E(X)$ even if nothing is known about the distribution of $X$.

**12.5** A fair coin is tossed independently $n$ times. Let $S_n$ be the number of heads obtained. Use the Chebyshev inequality to find a lower bound of the probability that $\frac{S_n}{n}$ differs from $\frac{1}{2}$ by less than 0.1 when a) $n = 100$, b) $n = 10{,}000$, and c) $n = 100{,}000$.

$\circ$ **12.6** Find Chebyshev's lower bound of the probability that $\frac{S_n}{n}$ in Problem 12.5 differs from $\frac{1}{2}$ by less than 0.01 when a) $n = 100$, b) $n = 10{,}000$, and c) $n = 100{,}000$.

The lower bounds obtained in Problems 12.5 and 12.6 seem to approach 1 as $n$ increases, even if 0.1 is replaced by any positive number $\varepsilon > 0$. If this is so, we have made an interesting observation: In a long series of independent coin tosses, the number of heads obtained divided by the total number of tosses is likely to be close to $\frac{1}{2}$, in fact arbitrarily close because $\varepsilon > 0$ can be as small as we wish. This is just what our intuition of probability tells us to expect. But now, it can take the form of a precise mathematical statement.

**12.7** In the same setting as in Problem 12.5, demonstrate that

$$\lim_{n\to\infty} P\left(\left\{\left|\frac{S_n}{n} - \frac{1}{2}\right| < \varepsilon\right\}\right) = 1$$

for any $\varepsilon > 0$.

**12.8** Consider an unfair coin with probability $p$ of heads. Let $S_n$ be the number of heads obtained when the coin is tossed independently $n$ times. Write down a limit similar to that in Problem 12.7 and prove it.

**12.9** We throw $n$ dice independently and add the points, denoting the sum by $S_n$. Extend the results in Problems 12.7 and 12.8 to this case.

The limits in Problems 12.7–12.9 can be found by means of the Chebyshev inequality without referring to the distribution of the random variable $S_n$. Although the distribution is well known in each of these cases, the point is not to use it in order to have a simple argument, and, more importantly, to be able to generalize it to cases in which the distribution is unknown.

**12.10** Prove the following *weak law of large numbers.*

► **Theorem 12.2** *Let $X_1, \ldots, X_n$ be independent identically distributed random variables with finite expectation and variance and let $S_n = X_1 + \cdots + X_n$. We set $a = E(X_i)$, which is the same for each $i$. Then,*

$$\lim_{n\to\infty} P\left(\left\{\left|\frac{S_n}{n} - a\right| < \varepsilon\right\}\right) = 1$$

*for any $\varepsilon > 0$.*

The law of large numbers is of great importance in probability theory. It provides a precise mathematical statement of our intuition that *the average value over a large number of independent realizations of a random variable $X$ is likely to be close to the expectation $E(X)$.* Moreover, when $X = \mathbb{I}_A$ is the indicator function of an event, the law of large numbers also supports our intuition of probability: *the average number of occurrences of an event $A$ in a series of many independent trials is likely to be close to the probability $P(A)$.*

The following *Khinchine weak law of large numbers* shows that the assumption about the existence and finiteness of variance can be relaxed (in which case the Chebyshev inequality will, of course, be of little use). The proof is omitted.

▶ **Theorem 12.3** *Let $X_1, \dots, X_n$ be independent identically distributed random variables with finite expectation $a = E(X_i)$ and let $S_n = X_1 + \dots + X_n$. Then,*

$$\lim_{n\to\infty} P\left(\left\{\left|\frac{S_n}{n} - a\right| < \varepsilon\right\}\right) = 1$$

*for any $\varepsilon > 0$.*

It is interesting that the weak law of large numbers implies the *Weierstrass theorem* on approximating continuous functions by polynomials. The polynomials below are called *Bernstein polynomials*.

* **12.11** Using the weak law of law of large numbers, prove that for every continuous function $f : [0,1] \to \mathbb{R}$,

$$\sum_{k=0}^{n} f\left(\frac{k}{n}\right)\binom{n}{k} x^k(1-x)^k \to f(x)$$

uniformly in $x \in [0,1]$ as $n \to \infty$.

An assertion stronger than the weak law of large numbers is also true. As it turns out, not only does the probability that $\frac{S_n}{n}$ is close to $a = E\left(\frac{S_n}{n}\right)$ tend to 1 as $n \to \infty$, but, in fact, $\frac{S_n}{n}$ itself converges to $a$ with probability 1. This is known as the *Kolmogorov strong law of large numbers* and is one of the major results in probability theory. Again, the proof is omitted.

▶ **Theorem 12.4** *Let $\{X_n\}$ be a sequence of independent identically distributed random variables and let $S_n = X_1 + \dots + X_n$. Then, the expectation $E(X_i) = a$ exists if and only if*

$$P\left(\left\{\lim_{n\to\infty} \frac{S_n}{n} = a\right\}\right) = 1.$$

The last theorem implies the weak law of large numbers because of the assertion in Problem 12.17. This is why it is called the "strong" law in contrast to the "weak" law.

**12.12** Let $S_n(x)$ denote the number of 1's among the first $n$ digits of the binary expansion of $x \in [0,1]$. Show that

$$\lim_{n\to\infty} \frac{S_n(x)}{n} = \frac{1}{2}$$

for all $x$'s in a set of Lebesgue measure 1 in $[0,1]$.

Suppose that a sample $X_1, \dots, X_n$ is obtained from a population to estimate the expectation $E(X)$ and variance $V(X)$ of a random variable $X$ so that the $X_i$ can be regarded as independent copies of $X$. It is natural to define an estimate of $E(X)$ to be

$$a_n = \frac{1}{n} \sum_{i=1}^{n} X_i.$$

We write $a_n$ instead of the standard notation $\bar{X}$ because explicit dependence on $n$ will be needed in what follows.

It also seems natural to take $\frac{1}{n} \sum_{i=1}^{n} (X_i - a_n)^2$ as an estimate of $V(X)$, but we shall see below that, in fact, this expression needs to be slightly modified.

**12.13** Show that if $X$ has finite expectation, then $E(a_n) = E(X)$ and

$$\lim_{n \to \infty} a_n = E(X)$$

with probability 1.

**12.14** Show that

$$E\left(\frac{1}{n} \sum_{i=1}^{n} (X_i - a_n)^2\right) = \frac{n-1}{n} V(X).$$

Because of this result, we define

$$\sigma_n^2 = \frac{1}{n-1} \sum_{i=1}^{n} (X_i - a_n)^2$$

to be an estimate of $V(X)$.

**12.15** Show that if $X$ has finite expectation and variance, then $E(\sigma_n^2) = V(X)$ and

$$\lim_{n \to \infty} \sigma_n^2 = V(X)$$

with probability 1.

**12.16** Let $(\Omega, \mathcal{F}, P)$ be a probability space, $\Omega$ being the set of all possible outcomes of a random experiment. Take two events $A, B \in \mathcal{F}$ with $P(B) \neq 0$. If the random experiment is repeated independently $n$ times, let $S_n$ be the number of times the outcome is in $B$. Among these $S_n$ outcomes in $B$, there will be $R_n$ outcomes that are also in $A$. Show that

$$\lim_{n \to \infty} \frac{R_n}{S_n} = P(A|B)$$

with probability 1.

In the weak and strong laws of large numbers, we have encountered two modes of convergence of random variables, namely convergence *in probability* and, respectively, *convergence with probability* 1, defined as follows:

▶ **Definition 12.1** A sequence of random variables $X_n$ is said to converge *in probability* to $X$ if

$$\lim_{n\to\infty} P\left(\{|X_n - X| < \varepsilon\}\right) = 1.$$

▶ **Definition 12.2** A sequence of random variables $X_n$ is said to converge *with probability* 1 (or *almost surely*) to a random variable $X$ if

$$P\left(\left\{\lim_{n\to\infty} X_n = X\right\}\right) = 1.$$

To state the central limit theorem (Theorem 12.5), we shall need yet another type of convergence.

▶ **Definition 12.3** We say that a sequence of random variables $X_n$ converges *in distribution* to $X$ if $F_{X_n}(x) \to F_X(x)$ as $n \to \infty$ for every point $x$ at which $F_X$ is continuous.

Here, $F_X$ denotes the distribution function of a random variable $X$; see Definition 8.3.

In the next two problems, we shall prove that convergence with probability 1 (almost surely) implies convergence in probability, which, in turn, implies convergence in distribution.

* **12.17** Prove that if a sequence of random variables $X_n$ converges almost surely to $X$, then it converges to $X$ in probability.

* **12.18** Prove that if a sequence of random variables $X_n$ converges to $X$ in probability, then it converges to $X$ in distribution.

According to the law of large numbers, if $S_n$ is the sum of $n$ independent identically distributed random variables $X_i$ with finite expectations $a = E(X_i)$, then $\frac{S_n}{n}$ converges almost surely to $E\left(\frac{S_n}{n}\right) = a$. The *Lindenberg-Lévy central limit theorem* below provides further information on the behavior of $S_n$ for large $n$. It is one of the great results of mathematics..

▶ **Theorem 12.5** *Let $\{X_n\}$ be a sequence of independent identically distributed random variables with finite expectation $a = E(X_n)$ and finite variance $\sigma^2 = V(X_n) > 0$. If $S_n = X_1 + \cdots + X_n$, then*

$$\lim_{n\to\infty} P\left(\left\{\frac{S_n - na}{\sigma\sqrt{n}} \leqslant x\right\}\right) = \frac{1}{\sqrt{2\pi}} \int_{-\infty}^{x} e^{-t^2/2}\,\mathrm{d}t;$$

*that is, $\frac{S_n - na}{\sigma\sqrt{n}}$ converges in distribution to a random variable having the normal distribution with mean 0 and variance 1.*

The proof is omitted. This theorem is illustrated in Figure 12.1. Here, the $X_n$ are independent random variables taking values 0 or 1 with probability $\frac{1}{2}$, so $a = E(X_n) = \frac{1}{2}$, $\sigma^2 = V(X_n) = \frac{1}{4}$, and $S_n = X_1 + \cdots + X_n$ is a random variable with binomial distribution. The step functions in Figure 12.1 have steps of width $\frac{1}{\sigma\sqrt{n}}$ such that the probability $P(\{S_n = k\}) = \binom{n}{k}2^{-n}$ is equal to the area under the step centered at $x = \frac{k-na}{\sigma\sqrt{n}}$. Letting $n \to \infty$, we obtain the density $\frac{1}{\sqrt{2\pi}}e^{-x^2/2}$ of the normal distribution.

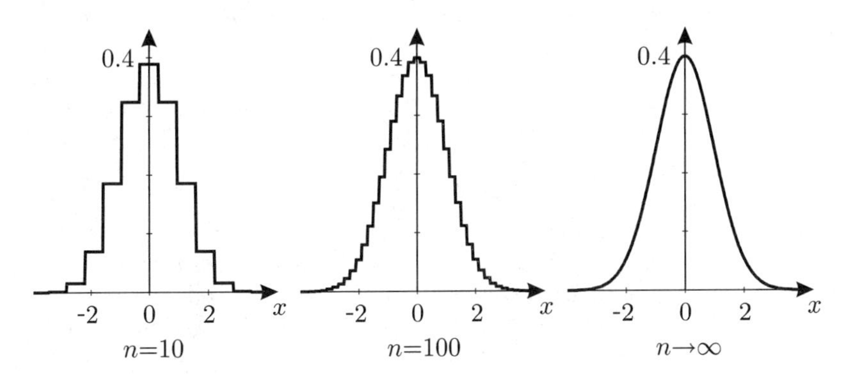


FIGURE 12.1. Step functions representing the distribution of $S_n$. The area under each step centered at $x = \frac{k-na}{\sigma\sqrt{n}}$ is equal to $P(\{S_n = k\})$. The normal density is obtained as $n \to \infty$.

The central limit theorem is sometimes expressed by saying that $\frac{S_n}{n}$ is asymptotically normally distributed with mean $E\left(\frac{S_n}{n}\right) = a$ and variance $V\left(\frac{S_n}{n}\right) = \frac{\sigma^2}{n}$, where $\sigma^2 = V(X_i)$, in the following sense.

**12.19** Demonstrate that for every real number $x$,

$$\lim_{n\to\infty} \frac{P\left(\left\{\frac{S_n}{n} \leqslant x\right\}\right)}{\frac{1}{\sqrt{2\pi(\sigma^2/n)}}\int_{-\infty}^{x} e^{-(t-a)^2/2(\sigma^2/n)}\,dt} = 1.$$

The problems below are of great importance because of practical applications, in which precise answers are often unnecessary. Approximations obtained with the aid of the central limit theorem are usually sufficient.

**12.20** Estimate the probability that the number of heads in 10,000 independent tosses of a fair coin differs by less than 1% from 5000.

What is the probability that the number of heads will be greater than 5100?

**12.21** How many independent tosses of a fair coin are required for the probability that the average number of heads differs from 0.5 by less than 1% to be at least 0.99?

**12.22** Three candidates $A$, $B$, and $C$ are running for President. In order to predict the outcome of the election, a number of people, each selected at random and independently of the others, are asked their choice. The predicted percentage $p_X$ of votes for a each candidate $X \in \{A, B, C\}$ is worked out by dividing the number of people in the sample who are going to vote for $X$ by the total number of people in the sample. How many people need to be in the sample to substantiate the following claim: *For each candidate $X$, the probability that the predicted percentage $p_X$ is correct to within 1% is at least* 0.95?

What sample size will make it possible to make the same claim if there are just two candidates? And when there are $n$ candidates? (Exclude $n = 1$, even though this case is not uncommon in certain political systems.)

**12.23** Suppose that a sample consisting of $n$ men is used in the survey described in Problem 7.22 to establish the likelihood of being a bathroom singer. Each person is asked to toss a die and answer YES or NO. The probability $P(Y)$ of answering YES is estimated by the average number $P_n Y$ of YES answers (that is, the number of YES answers divided by $n$). Because $P(Y)$ is connected with the probability $P(S)$ of being a bathroom singer by $P(Y) = \frac{1}{6} + \frac{2}{3}P(S)$ (see Solution 7.22), the estimate $P_n S$ of $P(S)$ is then computed from the formula

$$P_n Y = \frac{1}{6} + \frac{2}{3} P_n S.$$

Estimate the probability that $P_n S$ differs from $P(S)$ by more than $\varepsilon > 0$. Then, find how large a sample is needed to make sure that $P_n S$ differs from $P(S)$ by no more than 0.01 with probability at least 0.95? Compare this to the sample size needed when the survey question is answered directly without tossing a die.

## 12.2 Hints

**12.1** If $X$ is the distance between the point hit and the center of the target, consider the two random variables $X$ and $20\,\mathbb{I}_{\{X \geqslant 20\}}$, where $\mathbb{I}_A$ is the indicator function of a set $A$. Which of the two is greater? Consider the case when $X \geqslant 20$ and when $X < 20$, and take into account that $X$ is non-negative. Having compared the random variables, take expectations.

**12.2** Think of a random variable taking two values: 0 with probability $\frac{3}{4}$ and 20 with probability $\frac{1}{4}$. What is the expectation?

**12.3** Study Problem 12.1, replacing 20 by an arbitrary $C > 0$ and 5 by $E(X)$.

**12.4** Apply the inequality in Problem 12.3 to the random variable $Y = (X - E(X))^2$.

**12.5** What is the expectation $E\left(\frac{S_n}{n}\right)$? Now, apply the Chebyshev inequality to estimate the probability $P\left(\left\{\left|\frac{S_n}{n} - \frac{1}{2}\right| < 0.1\right\}\right)$. You will need to compute the variance $V\left(\frac{S_n}{n}\right)$. The independence of each toss can be of help when computing the expectation and variance. Or use the fact that $S_n$ has binomial distribution.

**12.6** See Problem 12.5.

**12.7** Just replace 0.1 in the solution to Problem 12.5 by $\varepsilon$ and then let $n$ tend to infinity.

**12.8** For an unfair coin, what do you expect the average number of heads $\frac{S_n}{n}$ to be close to when $n$ is large? What is $E\left(\frac{S_n}{n}\right)$? What is $V\left(\frac{S_n}{n}\right)$? To compute the expectation and variance, use the fact that the coin is tossed independently each time; see Solution 12.5. Now write down the Chebyshev inequality and let $n$ tend to infinity.

**12.9** What do you expect the mean number of points $\frac{S_n}{n}$ to be close to when $n$ is large? Now, proceed as in the solutions to Problems 12.7 and 12.8. In fact, there is no need to compute the variance of $\frac{S_n}{n}$, as long as you can show that $V\left(\frac{S_n}{n}\right) \to 0$ as $n \to \infty$.

**12.10** See Solutions 12.7–12.9.

**12.11** Let $S_n$ be the total number of heads when an unfair coin is tossed independently $n$ times. Consider $E\left(f\left(\frac{S_n}{n}\right)\right)$ as a function of the probability $p$ of heads. How does one apply the weak law of large numbers to estimate the difference between $E\left(f\left(\frac{S_n}{n}\right)\right)$ and $f(p)$, uniformly in $p$?

**12.12** Demonstrate that the digits in the binary expansion can be regarded as independent random variables under the Lebesgue measure on $[0, 1]$. What is their distribution? Use the strong law of large numbers to obtain the desired result.

**12.13** The assertion is equivalent to the strong law of large numbers, but expressed in a slightly different language.

**12.14** First, consider the case when $E(X) = 0$ for simplicity. Try to express $E\left(\frac{1}{n}\sum_{i=1}^{n}(X_i - a_n)^2\right)$ as the difference between $V(X)$ and $V(a_n)$. Then, compute $V(a_n)$.

**12.15** Because the random variables $(X_i - a_n)^2$, $i = 1, \ldots, n$, are not independent, in general, the law of large numbers for independent identically distributed random variables cannot be applied directly to the expression $\frac{1}{n-1}\sum_{i=1}^{n}(X_i - a_n)^2$ defining $\sigma_n^2$. It needs to be transformed so that independent random variables appear in the sum. For example, try to replace $(X_i - a_n)^2$ by $(X_i - E(X_i))^2$ and compute the remainder. In fact, this has already been done in Solution 12.14. Now, you will be able to apply the law of large numbers to the sum. The reminder is easily shown to converge to zero. The fact that we have $\frac{1}{n-1}$ rather than $\frac{1}{n}$ in front of the sum presents no difficulty.

**12.16** Take $Y = \mathbb{I}_B$ and represent $S_n$ as the sum of $n$ independent versions of $Y$. Similarly, represent $R_n$ as the sum of $n$ independent versions of a random variable. Then, apply the strong law of large numbers to find the limits of $\frac{S_n}{n}$ and $\frac{R_n}{n}$.

**12.17** You can assume that $X = 0$. First, prove that

$$\lim_{n\to\infty} P\left(\left\{\sup_{k\geqslant n}|X_k| < \varepsilon\right\}\right) = 1$$

for any $\varepsilon > 0$. This is a stronger assertion than convergence in probability, but, in fact, the supremum makes the task easier because the sets $\left\{\sup_{k\geqslant n}|X_k| < \varepsilon\right\}$ form an expanding sequence.

**12.18** Observe that $F_{X_n}(x) \leqslant F_X(x+\varepsilon) + P(\{|X_n - X| \geqslant \varepsilon\})$. Obtain a similar lower estimate for $F_{X_n}(x)$ in terms of $F_X(x-\varepsilon)$ and hence derive the assertion.

**12.19** Transform the inequality $\frac{S_n}{n} \leqslant x$ into $\frac{S_n - na}{\sigma\sqrt{n}} \leqslant \frac{x-a}{\sigma/\sqrt{n}}$ and then apply the central limit theorem.

**12.20** Consider the number of heads to be the sum of 10,000 independent copies of a random variable taking values 1 and 0 with probability $\frac{1}{2}$ and apply the central limit theorem.

**12.21** Apply the central limit theorem in a similar way as in Problem 12.20, but now for an arbitrary number $n$ of coin tosses. Then, find the smallest $n$ for which the resulting estimate of the probability is at least 0.99.

**12.22** The exact wording of the claim is important. It reads "for each candidate the probability that ..." rather than "the probability that for each candidate ... ." The latter would be a stronger statement, leading to a different and more difficult problem (and a larger minimum sample size, in general).

If $r_X$ denotes the probability that any given person will vote for candidate $X$, the claim means that $P(\{|p_X - r_X| \leqslant 0.01\}) \geqslant 0.95$ for each $X \in \{A, B, C\}$.

Fix $X$ and use the central limit theorem to estimate the size of the smallest sample for which $P(\{|p_X - r_X| \leqslant 0.01\}) \geqslant 0.95$. The estimate will depend on $r_X$. For which $r_X \in [0, 1]$ do you obtain the largest estimate? A sample of at least this size will have to be used because $r_X$ is unknown in advance (the aim of the survey being to determine the value of $r_X$).

Will a sample size estimated in this way for one candidate $X \in \{A, B, C\}$ be large enough for any other $X$? Does it matter that there are three candidates here, rather than just two or more than three?

**12.23** Write down the condition $|P_n S - P(S)| < \varepsilon$ in terms of $P_n Y$ and $P(Y)$. Then, express $P_n Y$ as the sum of $n$ independent copies of a random variable divided by $n$ and apply the central limit theorem.

If the survey question is answered directly without tossing a die, then a straightforward application of the central limit theorem to estimate the likelihood that $|P_n S - P(S)| < \varepsilon$, without writing the latter in terms of $P_n Y$ and $P(Y)$, is all that is needed.

## 12.3 Solutions

**12.1** Let $X$ be the distance between the point hit and the center of the target, measured in inches. Clearly, $X \geqslant 0$. Consider the random variable

$$20\,\mathbb{I}_{\{X\geqslant 20\}} = \begin{cases} 20 & \text{when } X \geqslant 20 \\ 0 & \text{when } X < 20. \end{cases}$$

It follows that

$$X \geqslant 20\,\mathbb{I}_{\{X\geqslant 20\}}.$$

Indeed, whenever $X < 20$, the right-hand side is 0, but $X$ is non-negative, so the inequality holds. When $X \geqslant 20$, the right-hand side is equal to 20 and the above inequality is also true.

Now, taking expectations, we obtain

$$E(X) \geqslant 20\,E(\mathbb{I}_{\{X\geqslant 20\}}) = 20\,P(\{X \geqslant 20\}).$$

But the left-hand side is just the average value, $E(X) = 5$, so that $P(\{X \geqslant 20\}) \leqslant \frac{1}{4}$.

**12.2** Let $X$ be a random variable taking two values: 0 with probability $\frac{3}{4}$ and 20 with probability $\frac{1}{4}$. The expected value is $E(X) = 5$. This is not a very realistic distribution of the distance between the point hit by an arrow and the center of the target, but it shows that $\frac{1}{4}$ is the best possible estimate in Problem 12.1.

A more realistic distribution (with density) can also be obtained for which the probability of missing the target is less than but arbitrarily close to $\frac{1}{4}$.

**12.3** Let $C > 0$. Since $X \geqslant C\,\mathbb{I}_{\{X \geqslant C\}}$,

$$E(X) \geqslant C\,E(\mathbb{I}_{\{X \geqslant C\}}) = C\,P(\{X \geqslant C\}),$$

which implies the estimate.

**12.4** By Problem 12.3, $P(\{Y \geqslant C^2\}) \leqslant \frac{E(Y)}{C^2}$; hence,

$$\begin{aligned} P(\{|X - E(X)| \geqslant C\}) &= P(\{(X - E(X))^2 \geqslant C^2\}) \\ &\leqslant \frac{E\left((X - E(X))^2\right)}{C^2} = \frac{V(X)}{C^2}. \end{aligned}$$

**12.5** Let $X_1, \ldots, X_n$ be independent random variables, each taking two values 0 and 1, both with probability $\frac{1}{2}$. This clearly corresponds to tossing a fair coin independently $n$ times. Let us say that 1 stands for heads and 0 for tails. Then, $S_n = X_1 + \cdots + X_n$ will be the number of heads obtained. We find that

$$E\left(\frac{S_n}{n}\right) = \frac{1}{n}\sum_{i=1}^{n} E(X_i) = \frac{1}{2}$$

and, because $X_1, \ldots, X_n$ are independent,

$$V\left(\frac{S_n}{n}\right) = \frac{1}{n^2}\sum_{i=1}^{n} V(X_i) = \frac{1}{4n},$$

since $E(X_i) = \frac{1}{2}$ and $V(X_i) = \frac{1}{4}$ for any $i \in \{1, \ldots, n\}$. Thus, by the Chebyshev inequality,

$$P\left(\left\{\left|\frac{S_n}{n} - \frac{1}{2}\right| \geqslant 0.1\right\}\right) \leqslant \frac{1}{4(0.1)^2 n}.$$

This means that

$$P\left(\left\{\left|\frac{S_n}{n} - \frac{1}{2}\right| < 0.1\right\}\right) \geqslant 1 - \frac{1}{4(0.1)^2 n}.$$

Substituting the values of $n$ in cases a), b), and c), we find the lower bound on the right-hand side to be $1 - \frac{1}{4}$, $1 - \frac{1}{400}$, and $1 - \frac{1}{40{,}000}$, respectively.

**12.7** As in the solution of Problem 12.5, we can show that

$$P\left(\left\{\left|\frac{S_n}{n} - \frac{1}{2}\right| < \varepsilon\right\}\right) \geqslant 1 - \frac{1}{4\varepsilon^2 n},$$

which means that the probability tends to 1 as $n \to \infty$, for any given $\varepsilon > 0$.

**12.8** Let $X_1, \dots, X_n$ be independent random variables, each taking two values: 1 with probability $p$ and 0 with probability $1-p$, so that 1 stands for heads and 0 for tails. Then, $S_n = X_1 + \cdots + X_n$ will be the number of heads obtained. We find that

$$E\left(\frac{S_n}{n}\right) = \frac{1}{n}E(S_n) = \frac{1}{n}\sum_{i=1}^{n} E(X_i) = p$$

and, because $X_1, \dots, X_n$ are independent,

$$V\left(\frac{S_n}{n}\right) = \frac{1}{n^2}V(S_n) = \frac{1}{n^2}\sum_{i=1}^{n} V(X_i) = \frac{p(1-p)}{n},$$

since $E(X_i) = p$ and $V(X_i) = p(p-1)$ for each $i \in \{0, \dots, n\}$. Thus, by the Chebyshev inequality,

$$P\left(\left\{\left|\frac{S_n}{n} - p\right| \geqslant \varepsilon\right\}\right) \leqslant \frac{p(p-1)}{\varepsilon^2 n}$$

for any $\varepsilon > 0$. This implies that

$$\lim_{n\to\infty} P\left(\left\{\left|\frac{S_n}{n} - p\right| < \varepsilon\right\}\right) = 1.$$

**12.9** Let $X_i$ be the number of points shown by the $i$th die. Then, $S_n = X_1 + \cdots + X_n$. Because the $X_i$ are identically distributed, we find that

$$E\left(\frac{S_n}{n}\right) = \frac{1}{n}\sum_{i=1}^{n} E(X_i) = E(X_1) = 3.5,$$

which is what we expect $\frac{S_n}{n}$ to be close to when $n$ is large. Furthermore, since the $X_i$ are independent,

$$V\left(\frac{S_n}{n}\right) = \frac{1}{n^2}\sum_{i=1}^{n} V(X_i) = \frac{1}{n}V(X_1) \to 0$$

as $n \to \infty$. By the Chebyshev inequality,

$$P\left(\left\{\left|\frac{S_n}{n} - E\left(\frac{S_n}{n}\right)\right| < \varepsilon\right\}\right) \geqslant 1 - \frac{V\left(\frac{S_n}{n}\right)}{\varepsilon^2}$$

for any $\varepsilon > 0$. It follows that

$$\lim_{n\to\infty} P\left(\left\{\left|\frac{S_n}{n} - 3.5\right| < \varepsilon\right\}\right) = 1$$

for any $\varepsilon > 0$. This extends the results in Problems 12.7 and 12.8 to the case in hand.

**12.10** The argument is the same as in Solution 12.9. Just replace 3.5 by $a$.

**12.11** Let $X_1, \ldots, X_n$ and $S_n$ be as in Solution 12.8 on tossing a coin with probability $p$ of heads. We put

$$B_n(p) = E\left(f\left(\frac{S_n}{n}\right)\right) = \sum_{k=0}^{n} f\left(\frac{k}{n}\right)\binom{n}{k} p^k(1-p)^k.$$

Then,

$$\begin{aligned} |B_n(p) - f(p)| &= \left|E\left(f\left(\frac{S_n}{n}\right)\right) - f(p)\right| \\ &\leqslant E\left|f\left(\frac{S_n}{n}\right) - f(p)\right| = E(Y_n), \end{aligned}$$

where $Y_n = \left|f\left(\frac{S_n}{n}\right) - f(p)\right|$. We write $A_n^\varepsilon = \left\{\left|\frac{S_n}{n} - p\right| < \varepsilon\right\}$ for any $\varepsilon > 0$. Then,

$$\begin{aligned} E(Y_n) &= E(Y_n \mathbb{I}_{A_n^\varepsilon}) + E(Y_n \mathbb{I}_{\Omega \setminus A_n^\varepsilon}) \\ &\leqslant \sup_{|x|<\varepsilon} |f(p+x) - f(p)| + 2 \sup_{x\in[0,1]} |f(x)|\, P(\Omega \setminus A_n^\varepsilon). \end{aligned}$$

For any $\eta > 0$, the first term on the right-hand side can be made smaller than $\frac{\eta}{2}$ by choosing a sufficiently small $\varepsilon > 0$. Then, by the weak law of large numbers, the second term will also be smaller than $\frac{\eta}{2}$ if $n$ is large enough. It follows that $|B_n(p) - f(p)| \leqslant \eta$ for every $p \in [0, 1]$ if $n$ is large enough, which completes the proof.

**12.12** Let $\{X_n\}$ be a sequence of independent random variables, each being equal to 1 or 0 with probability $\frac{1}{2}$. We take

$$Y = \sum_{n=1}^{\infty} \frac{X_n}{2^n}.$$

This is a random variable with values in the unit interval $[0, 1]$. It is well defined because the series is always convergent. The $X_n$ are the digits in the binary expansion of $Y$.

We claim that $m(A) = P(\{Y \in A\})$, where $A \subset [0,1]$ is any Borel set, is the Lebesgue measure on $[0,1]$. Indeed, for any $y \in [0,1]$ with binary expansion $y = \sum_{n=1}^{\infty} \frac{x_n}{2^n}$,

$$\{Y < y\} = \bigcup_{n=1}^{\infty} \{X_1 = x_1, \ldots, X_{n-1} = x_{n-1}, X_n < x_n\}.$$

Thus, by the independence of the $X_n$,

$$\begin{aligned} P(\{Y < y\}) &= \sum_{n=1}^{\infty} \left(\frac{1}{2}\right)^n P(\{X_n < x_n\}) \\ &= \sum_{n=1}^{\infty} \frac{x_n}{2^n} = y = m([0,y)). \end{aligned}$$

*Caution:* This argument breaks down for any $y$ that has a finite binary expansion (for example, $\frac{1}{4} = 0.01$). The reason is that there is also an infinite binary expansion for the same $y$ (for example, $\frac{1}{4} = 0.00111\ldots$). But the set of such $y$'s is of Lebesgue measure zero because it is countable (see Chapter 6), so it does not affect our conclusion.

Now, because $S_n(Y) = X_1 + \cdots + X_n$, by the strong law of large numbers it follows that

$$\begin{aligned} m\left(\left\{x \in [0,1] : \lim_{n\to\infty} \frac{S_n(x)}{n} = \frac{1}{2}\right\}\right) &= P\left(\left\{\lim_{n\to\infty} \frac{S_n(Y)}{n} = \frac{1}{2}\right\}\right) \\ &= 1, \end{aligned}$$

as required.

**12.13** The proof that the expectations are equal is straightforward:

$$E(a_n) = \frac{1}{n}\sum_{i=1}^{n} E(X_i) = \frac{1}{n}\sum_{i=1}^{n} E(X) = E(X).$$

The convergence $a_n \to E(X)$ with probability 1 is precisely what is asserted in the strong law of large numbers.

**12.14** We observe that

$$\begin{aligned} E(a_n) &= E(X), \\ V(a_n) &= \frac{1}{n} V(X). \end{aligned}$$

The former equality has been verified in Problem 12.13. The latter follows by the independence of $X_1, \ldots, X_n$, namely

$$V(a_n) = V\left(\frac{1}{n}\sum_{i=1}^{n} X_i\right) = \frac{1}{n^2}\sum_{i=1}^{n} V(X_i) = \frac{1}{n} V(X).$$

Now consider the case when $E(X) = 0$ for simplicity. Then,

$$\frac{1}{n}\sum_{i=1}^{n}(X_i - a_n)^2 = \frac{1}{n}\sum_{i=1}^{n}\left(X_i^2 - 2X_i a_n + (a_n)^2\right)$$
$$= \frac{1}{n}\sum_{i=1}^{n} X_i^2 - (a_n)^2.$$

The general case when $E(X)$ is not necessarily zero can be obtained from the above by replacing $X$ by $X - E(X)$ and, correspondingly, $X_i$ by $X_i - E(X_i)$, leading to

$$\frac{1}{n}\sum_{i=1}^{n}(X_i - a_n)^2 = \frac{1}{n}\sum_{i=1}^{n}(X_i - E(X_i))^2 - (a_n - E(a_n))^2.$$

Next, taking the expectation, we obtain

$$E\left(\frac{1}{n}\sum_{i=1}^{n}(X_i - a_n)^2\right) = V(X) - V(a_n) = \frac{n-1}{n}V(X),$$

as required.

**12.15** From the solution of Problem 12.14 it follows immediately that

$$\sigma_n^2 = \frac{n}{n-1}\frac{1}{n}\sum_{i=1}^{n}(X_i - E(X_i))^2 - \frac{n}{n-1}(a_n - E(a_n))^2.$$

In Problem 12.13, we have seen that $a_n \to E(X) = E(a_n)$ with probability 1, which means that the second term on the right-hand side converges to zero with probability 1. In the first term, the summands $(X_i - E(X_i))^2$ are independent identically distributed random variables with finite expectation $E\left((X_i - E(X_i))^2\right) = V(X)$. Therefore, by the strong law of large numbers,

$$\lim_{n\to\infty}\frac{1}{n}\sum_{i=1}^{n}(X_i - E(X_i))^2 = V(X)$$

with probability 1, which implies that $\sigma_n^2 \to V(X)$ with probability 1 because $\frac{n}{n-1} \to 1$ as $n \to \infty$.

**12.16** Let $\omega_1, \ldots, \omega_n \in \Omega$ be the outcomes of $n$ independent versions of the random experiment. We put

$$X_i = \mathbb{I}_{A\cap B}(\omega_i), \qquad Y_i = \mathbb{I}_B(\omega_i)$$

for $i = 1, \ldots, n$. Then,

$$R_n = X_1 + \cdots + X_n, \qquad S_n = Y_1 + \cdots + Y_n.$$

The $X_i$ are independent random variables and so are the $Y_i$. By the strong law of large numbers,

$$\lim_{n\to\infty} \frac{R_n}{n} = E\left(\mathbb{I}_{A\cap B}\right) = P(A\cap B), \qquad \lim_{n\to\infty} \frac{S_n}{n} = E\left(\mathbb{I}_{B}\right) = P(B)$$

with probability 1. Since $P(B) \neq 0$,

$$\lim_{n\to\infty} \frac{R_n}{S_n} = \lim_{n\to\infty} \frac{R_n}{n}\frac{n}{S_n} = \frac{P(A\cap B)}{P(B)} = P(A|B)$$

with probability 1.

**12.17** We can assume without loss of generality that $X = 0$, for otherwise we could take $X_n - X$ in place of $X_n$. Fix any $\varepsilon > 0$ and put

$$A_n^\varepsilon = \{|X_n| < \varepsilon\},$$
$$B_n^\varepsilon = \left\{\sup_{k\geqslant n} |X_k| < \varepsilon\right\}.$$

Because

$$\left\{\lim_{n\to\infty} X_n = 0\right\} \subset \bigcup_{n=1}^{\infty} B_n^\varepsilon,$$

$P\left(\{\lim_{n\to\infty} X_n = 0\}\right) = 1$ by assumption, and $B_n^\varepsilon \subset B_{n+1}^\varepsilon$, it follows that

$$1 = P\left(\bigcup_{n=1}^{\infty} B_n^\varepsilon\right) = \lim_{n\to\infty} P(B_n^\varepsilon);$$

see Problem 6.8. But $B_n^\varepsilon \subset A_n^\varepsilon$, so that $\lim_{n\to\infty} P(A_n^\varepsilon) = 1$, completing the proof.

**12.18** Let $\varepsilon > 0$ and let $x$ be any real number. Because $X \leqslant X_n + |X_n - X|$, it follows that if $X_n \leqslant x$, then $X \leqslant x + \varepsilon$ or $|X_n - X| \geqslant \varepsilon$. This means that

$$P\left(\{X_n \leqslant x\}\right) \leqslant P\left(\{X \leqslant x + \varepsilon\}\right) + P\left(\{|X_n - X| \geqslant \varepsilon\}\right).$$

Similarly, we find that

$$P\left(\{X \leqslant x - \varepsilon\}\right) \leqslant P\left(\{X_n \leqslant x\}\right) + P\left(\{|X_n - X| \geqslant \varepsilon\}\right).$$

Letting $n \to \infty$, we obtain

$$\begin{aligned} P\left(\{X \leqslant x - \varepsilon\}\right) &\leqslant \liminf_{n\to\infty} P\left(\{X_n \leqslant x\}\right) \\ &\leqslant \limsup_{n\to\infty} P\left(\{X_n \leqslant x\}\right) \leqslant P\left(\{X \leqslant x + \varepsilon\}\right), \end{aligned}$$

since $P(\{|X_n - X| \geqslant \varepsilon\}) \to 0$ as $n \to \infty$ by assumption.

Whenever the distribution function $F_X(x) = P(\{X \leqslant x\})$ is continuous at $x$, $P(\{X \leqslant x - \varepsilon\})$ and $P(\{X \leqslant x + \varepsilon\})$ both tend to $F_X(x)$ as $\varepsilon \to 0$, proving the assertion.

**12.19** Because

$$\left\{\frac{S_n}{n} \leqslant x\right\} = \left\{\frac{S_n - na}{\sigma\sqrt{n}} \leqslant \frac{x-a}{\sigma/\sqrt{n}}\right\},$$

by the central limit theorem we find that

$$\lim_{n\to\infty} \frac{P\left(\{\frac{S_n}{n} \leqslant x\}\right)}{\frac{1}{\sqrt{2\pi}}\int_{-\infty}^{\frac{x-a}{\sigma/\sqrt{n}}} e^{-t^2/2}\,dt} = 1.$$

We can complete the argument by changing the integration variable to $t'$ such that $t = \frac{t'-a}{\sigma/\sqrt{n}}$.

**12.20** Let $S$ be the number of heads in 10,000 independent tosses of a fair coin. Then, $S$ can be considered as the sum of 10,000 independent copies of a random variable $X$ taking values 1 and 0 with probability $\frac{1}{2}$. The variance of $X$ is $\frac{1}{4}$. The central limit theorem yields

$$\begin{aligned} P(\{|S - 5000| < 50\}) &= P\left(\left\{\frac{|S-5000|}{\sqrt{10{,}000/4}} < 1\right\}\right) \\ &\approx \frac{1}{\sqrt{2\pi}}\int_{-1}^{+1} e^{-t^2/2}dt \approx 0.68. \end{aligned}$$

Similarly,

$$\begin{aligned} P(\{5100 < S\}) &= P\left(\left\{2 < \frac{S-5000}{\sqrt{10{,}000/4}}\right\}\right) \\ &\approx \frac{1}{\sqrt{2\pi}}\int_{2}^{\infty} e^{-t^2/2}dt \approx 0.023. \end{aligned}$$

**12.21** Let $S_n$ be the number of heads among $n$ tosses of a coin. By the central limit theorem,

$$\begin{aligned} P\left(\left\{\left|\frac{S_n}{n} - 0.5\right| < 0.005\right\}\right) &= P\left(\left\{\left|\frac{S_n - 0.5n}{\sqrt{n/4}}\right| < 0.01\sqrt{n}\right\}\right) \\ &\approx \frac{1}{\sqrt{2\pi}}\int_{-0.01\sqrt{n}}^{+0.01\sqrt{n}} e^{-t^2/2}dt \geqslant 0.99. \end{aligned}$$

This will be satisfied if $0.01\sqrt{n} \geqslant 2.59$; that is, $n \geqslant 259^2 = 67{,}081$.

**12.22** We need to estimate the smallest number $n$ of people in the sample such that $P(\{|p_X - r_X| \leqslant 0.01\}) \geqslant 0.95$ for any $r_X \in [0,1]$ for each candidate $X \in \{A, B, C\}$ (see the hint).

Let $S_X$ be the number of people in the sample who are going to vote for candidate $X$. Assuming that each person's preference is independent of the others, we can consider $S_X$ to be the sum of $n$ independent random variables, each taking value 1 with probability $r_X$ if a person will vote for candidate $X$ or 0 otherwise. The variance of each of these independent random variables is therefore $r_X(1 - r_X)$. Thus, by the central limit theorem,

$$P(\{|p_X - r_X| \leqslant 0.01\}) = P\left(\left\{\frac{|S_X - nr_X|}{\sqrt{nr_X(1-r_X)}} \leqslant 0.01\sqrt{\frac{n}{r_X(1-r_X)}}\right\}\right)$$

$$\approx \frac{1}{\sqrt{2\pi}} \int_{-0.01\sqrt{\frac{n}{r_X(1-r_X)}}}^{+0.01\sqrt{\frac{n}{r_X(1-r_X)}}} e^{-t^2/2} dt \geqslant 0.95.$$

This will be satisfied if

$$0.01\sqrt{\frac{n}{r_X(1-r_X)}} \geqslant 2.$$

In fact, we could use about 1.96 here instead of 2, but there is no need for such precision.

Because for $r_X \in [0,1]$ the largest value of $r_X(1 - r_X)$ is $\frac{1}{4}$, we find that the latter estimate will be satisfied for all $r_X \in [0,1]$ if

$$n \geqslant 10{,}000.$$

The same value will be obtained for each candidate $X \in \{A, B, C\}$.

Therefore, a sample of about 10,000 people is sufficient to claim that the predicted percentage of votes is accurate to within 1% with probability at least 0.95 for each candidate. The answer is the same for any number of candidates greater than one.

**12.23** Let $X$ be a random variable taking two values: 1 with probability $P(Y)$ and 0 with probability $1 - P(Y)$, so that it has variance $V(X) = P(Y)(1 - P(Y))$. Take $T_n$ be the sum of $n$ independent copies of $X$. Then, $P_n Y = \frac{T_n}{n}$.

Since $P(Y) = \frac{1}{6} + \frac{2}{3}P(S)$ and $P_nY = \frac{1}{6} + \frac{2}{3}P_nS$,

$$
\begin{aligned}
P\left(\{|P_nS - P(S)| < \varepsilon\}\right) &= P\left(\left\{|P_nY - P(Y)| < \frac{2}{3}\varepsilon\right\}\right) \\
&= P\left(\left\{\frac{|T_n - nP(Y)|}{\sqrt{nV(X)}} < \frac{2}{3}\varepsilon\sqrt{\frac{n}{V(X)}}\right\}\right) \\
&\approx \frac{1}{\sqrt{2\pi}}\int_{-\frac{2}{3}\varepsilon\sqrt{\frac{n}{V(X)}}}^{+\frac{2}{3}\varepsilon\sqrt{\frac{n}{V(X)}}} e^{-t^2/2}\mathrm{d}t \\
&\geqslant \frac{1}{\sqrt{2\pi}}\int_{-\frac{4}{3}\varepsilon\sqrt{n}}^{+\frac{4}{3}\varepsilon\sqrt{n}} e^{-t^2/2}\mathrm{d}t.
\end{aligned}
$$

The last inequality holds because the maximum possible value of $V(X) = P(Y)\,(1 - P(Y))$ is $\frac{1}{4}$.

In particular, setting $\varepsilon = 0.01$, we find that

$$
\frac{1}{\sqrt{2\pi}}\int_{-\frac{0.04}{3}\sqrt{n}}^{+\frac{0.04}{3}\sqrt{n}} e^{-t^2/2}\mathrm{d}t \geqslant 0.95
$$

if $\frac{0{,}04}{3}\sqrt{n} \geqslant 2$, i.e., if $n \geqslant 22{,}500$. A sample of at least this size must used to meet the condition specified in the problem.

If the survey question is answered directly without tossing a die, then $\frac{4}{3}\varepsilon\sqrt{n}$ must be replaced by $2\varepsilon\sqrt{n}$ in the integration limits above, which means that the minimum sample size is $\left(\frac{2}{3}\right)^2$ times the last one, i.e., about 10,000 people.

# Bibliography

Ash, R. B. (1970), *Basic Probability Theory*, John Wiley & Sons, New York.

Ash, R. B. (1972), *Real Analysis and Probability*, Academic Press, New York.

Billingsley, P. (1986), *Probability and Measure*, 2nd edition, John Wiley & Sons, New York.

Brzeźniak, Z. and Zastawniak, T. (1999), *Basic Stochastic Processes*, Springer-Verlag, London.

Capiński, M. and Kopp, E. (1999), *Measure, Integral and Probability*, Springer-Verlag, London.

Chung, K. L. (1974), *A Course in Probability Theory*, Academic Press, New York.

Dudley, R. M. (1989), *Real Analysis and Probability*, Wadsworth & Brooks-Cole, Pacific Grove, CA.

Grimmet, G. R and Stirzaker, D. R. (1992), *Probability and Random Processes*, 2nd edition, Clarendon Press, Oxford.

Grimmet, G. R and Welsh, D. J. A. (1986), *Probability, an Introduction*, Clarendon Press, Oxford.

McColl, J. H. (1995), *Probability*, Edward Arnold, London.

Ross, S. (1976), *A First Course in Probability*, Macmillan, New York.

Rudin, W. (1987), *Real and Complex Analysis*, 3rd edition, McGraw-Hill, New York.

Stirzaker, D. (1999), *Probability and Random Variables*, Cambridge University Press, Cambridge.

Williams, D. (1991), *Probability with Martingales*, Cambridge University Press, Cambridge.

# Index

## Problem Books in Mathematics *(continued)*

**An Outline of Set Theory**
by *James M. Henle*

**Demography Through Problems**
by *Nathan Keyfitz and John A. Beekman*

**Theorems and Problems in Functional Analysis**
by *A.A. Kirillov and A.D. Gvishiani*

**Exercises in Classical Ring Theory**
by *T.Y. Lam*

**Problem-Solving Through Problems**
by *Loren C. Larson*

**Winning Solutions**
by *Edward Lozansky and Cecil Rosseau*

**A Problem Seminar**
by *Donald J. Newman*

**Exercises in Number Theory**
by *D.P. Parent*

**Contests in Higher Mathematics:**
**Miklós Schweitzer Competitions 1962-1991**
by *Gábor J. Székely (editor)*